Our Uncertain Future

When Digital Evolution, Global Warming and Automation Converge

David Mills, Ph.D.

David Mills, "Our Uncertain Future: When Digital Evolution, Global Warming and Automation Converge," Second Edition, Amazon Books (2019).

First printing: September, 2013.
Updated/Second printing: November,
2013. Third printing: January, 2014.
Second edition: April, 2019.

For additional copies or information on different formats available, or for questions, feedback or comments to the author:

DavidMillsPhD@myfastmail.com

ISBN-10: 057813036X

ISBN-13: 978-0-578-13036-1

Key Words: future computers, climate change, digital age, intelligent machines, automation impact.

Many thanks to Steve Janzen for the cover design.

ACKNOWLEGEMENTS

Thanks to Marilyn Wilson for assistance throughout this project, for invaluable service as a sounding board, clipping service and editor. Thanks also to Dennis McKee, Janet Mills, Michael Mills and anonymous readers for valuable comments on earlier drafts. I also acknowledge considerable debt to the large number of experts and journalists, listed in the Bibliography and End Notes, whose excellent works formed the background for this work.

This book is dedicated to my son, Michael, who gave me one of the best reasons to undertake a serious study of the future.

This second printing incorporates editorial changes suggested by early readers, including moving detailed numerical results and graphs to an appendix and adding chapter summaries. The book was also updated by incorporating important developments since the first printing - which was only two months ago! *A book about the future can never be "finished"* - especially when the future is arriving as fast as it is now.

Five years later: Nothing could be more true than this last statement above! **This second edition** incorporates important changes and trends that have become evident in the last five years: the trends are summarized in the **Preface to the Second Edition**, and the resulting changes in predictions in a new section, **Conclusions Five Years Later**. The book has otherwise been left unchanged because, while the emphasis on some topics might be changed, the arguments and examples in the first edition remain largely valid. Readers may judge for themselves.

TABLE OF CONTENTS

Graphs and Figures

Tables

PREFACE TO THE SECOND EDITION

The first edition (2014) focused on three related, important questions regarding the future of the human race, issues raised by a number of different writers and thinkers. (See first edition for all references to sources.)

> Will the increase in computer intelligence lead to computers whose capabilities vastly exceed human intelligence, and will this outcome be beneficial or detrimental to human society?
>
> Will the worldwide implementation of automation lead to large scale civil disorder due to angry multitudes of starving former workers?
>
> Will human civilization succumb to the cumulative effects of global climate change, as the carbon dioxide level continues to rise relentlessly due to our inability to constrain ourselves?

But more important than these individual questions is the core issue considered by the present book, and few (if any) others: What happens when all three of these trends occur at the same time? Which dominates, what determines what the major effects will be on human existence?

In the first edition, to consider these important questions all the data available up to and including the year 2013 was considered and extrapolated into the future to about 2100. The conclusion I reached, *assuming the present trends held*, was that the dominant factor by far would be the advance of super-intelligent machines. Increasing far faster than such slower changes such as climate change or the spread of automation, computer speeds, by doubling every year, would well exceed that required to produce a machine of super-human intelligence at least by the year 2035. It was argued that, once operational, the existence of such an entity would propel us into a new future, one in which the human intelligence was no longer capable of foreseeing or predicting. In this sense, our future is expected to pass through a singularity, after which we can no longer meaningfully project.

It might be, for example, that such an intelligence would lead to solutions to the global climate change problem far outside what we can now see. It might be, as another example, that the advent of such computers will provide for activities and entertainments that will satisfactorily occupy the hordes no longer traditionally employed. (We already see the beginnings of this in streaming services, for example.) We cannot know whether such positive benefits will accrue, or that the new machines will instead be indifferent or hostile to human endeavors.

The key phrase in the above is, however, "*assuming the present trends held.*" It appears, only five years later, that there are several possible changes in the trend lines that could be important for the resulting predictions.

"Death" of Moore's law and the development of super-intelligent computers

The date of arrival of super-human computer intelligence is now somewhat more in doubt as computer speeds *appear* not to be rising as fast as they had been. For this application, the important speed is that of the fastest supercomputer on the planet. Moore's law essentially was the prediction that such speeds would continue to increase by doubling once a year. In Figure 4, observed speeds up to 2013 led to a prediction (dashed line) that the speed of the fastest supercomputers would double every year, to reach a speed exceeding 10^{23} FLOPS (FLoating point Operations Per Second) by the year 2035. This is approximately the speed required to enable supercomputers to reach super-human intelligence levels. While it is too soon to be sure, the relatively few data points since 2013 look like the doubling time may have increased to about two years. If this is true, and computer speeds continue to increase exponentially but with a doubling time twice that previous, the supercomputer speeds would greatly exceed human level intelligence more than twenty years later, in 2056 instead of 2035. Because of the extreme (and exponential) extrapolation required, this estimate must be considered the most tentative of any of those considered in this book.

In this connection, one should at least mention the possibility that there could be that a hard limit to supercomputer speed that might be encountered, so that the level of super-human intelligence is never reached. This seems unlikely, however, since massive parallelism in computer structure can compensate for lack of speed in individual processors, much as the human brain achieves human-level intelligence in spite of the intrinsic slowness of neuron speeds, through its use of massive parallelism in structure. It also seems true that, given the important applications that require increased computer speed, innovative ways to build faster -- or at least more capable --computers will ultimately be found. Another possibility is that the slowdown in increase in supercomputer speed is only a momentary "hiccup," and that a technological breakthrough will allow speeds to soon return to increasing at a rate in which the speed doubles every year. A final consideration is that programming counts, too. The estimate of required speed being 10^{23} FLOPS is based on the speed required to exceed the estimated computational capacity of the human brain by a factor of 10,000. It is likely that clever programing and design structure will reduce this requirement considerably.

In summary, the statement in the first edition that the fastest computers would probably reach speeds allowing super-human intelligence by the year 2035 should be replaced by the probability that such speeds will be reached somewhere between 2035 and 2056.

Effects of global climate change may be increasing faster than expected.

In the last five years, there have been a number of extreme weather events that have apparently been made more extreme by the effects of global warming. Such attributions can only be made by using supercomputers to model the weather with and without the temperature increase due to global warming. Because a warmer atmosphere holds more water vapor, it leads to storms of higher intensity and extent than otherwise would occur: Both severe storms with more rain and cold conditions are expected as well as heat waves and droughts. The major concern is that the planet is heading toward a tipping point after which the effects of climate change would be drastically (nonlinearly or catastrophically) increased. Whether or not this occurs anytime soon, it is safe to say that the world's countries have not managed to slow down carbon dioxide production sufficiently to avoid major problems in the future: We already seem to be "committed" to a rise of at least 2 °C.

Changes in issues involving automation and displaced workers.

It has become clear recently that the effect of automation is not to simply replace workers, but that automation (and artificial intelligence) will allow (and to some extent drive) the invention of completely new jobs that the displaced workers can migrate to. For example, it can be seen that a large part of what is called the "sharing" economy actually depends on automation/AI to exist. Ride sharing, for example, relies on an automatic program hosted on the internet to connect riders and drivers and to transfer payments, and to provide instant feedback to the driver's future passengers, as well as a GPS enabled map and driving system essential for its function. While there are certainly problems with the companies tending to pay the minimum survivable wage to the drivers, these services provide drivers with an opportunity to make money and riders an opportunity to save money. Other services, such as the selling opportunities for small businesses offered by Amazon and eBay, including the self-publishing services of Amazon, offer other ways to be gainfully employed, or at least to keep busy while making a little money. Automation is also responsible for much of the production and consumption of entertainment today, such as streaming services providing movie content, music, etc.

The suggestion that automation may create as many jobs as it destroys is supported by the fact that the U.S. labor force currently shows record low levels of unemployment *at the same time* as job loss due to automation and outsourcing proceeds apace. That automation, nonetheless, has disruptive effects is evidenced by the finding that (in the U.S. at least) the created jobs generally pay less, and have fewer benefits, than the jobs that are destroyed. The result is that the gap between rich and poor has gotten wider in the US in the last five years. Under the Trump administration, unfortunately, the tax structure – which might otherwise had offered some additional support to poor workers -- instead is making the problem worse by giving tax relief to the rich rather than the poor, and potentially cutting social programs as a consequence.

Given the changes that are occurring, it is looking less likely that some scheme of a direct, guaranteed income (as suggested in the first edition) will actually be required to offset the negative effects of automation, nor does it seem politically feasible as long as unemployment is low. Rather, it now seems more likely that there will continue to evolve alternative ways of occupying one's time while making enough money to at least get by, many of them enabled by computer programs (automation). Social programs will probably be implemented, where possible, to give displaced workers further support, for example, through minimum wage laws, improving the tax structure (possibly including a negative income tax) and social programs such as health insurance and college tuition support which primarily benefit the workers in the lower income brackets. All of these suggestions have support from the majority of the current Democratic candidates for president.

The consequences of the above for the predictions made in the first edition are summarized in a new section, **Conclusions Five Years Later**.

1. OVERVIEW: WHAT'S IN THIS BOOK

> There are many important problems currently facing the human race, forces pulling us in many different directions. Provided our society gives adequate responses, and considering the interactions of all these forces, it seems likely that by 2050 most of our current problems will be on the way to being solved. A major contribution to these solutions is expected to come from the continued development of digital technology, including by 2035 general machine intelligence exceeding human intelligence. The almost inevitable development of machine superintelligence following this stage will provide new challenges and opportunities for the human race.

What is in our future? Will it be a climate catastrophe, massive unemployment and economic disaster, robots and superintelligent machines that make us obsolete or eliminate us? Or something else?

Many different predictions about our future have appeared recently, most mutually exclusive.

- Some climate scientists and environmentalists predict a dismal future for apparently valid reasons.[1]
- Some experts in digital technology predict glorious futures of different kinds.[2]
- Experts in other areas point to the future glories of nano technology or genetic manipulation, including extension of the human lifespan and the end of disease.[3]
- Other analysts predict massive problems will come from the same technologies.[4]
- Some experts predict that our economy will fail as automation and other robots take all our jobs.[5]
- Other experts argue that we will be overcome by resource depletion and famine.[6]
- Finally, some worry that superintelligent machines may destroy the human race or at the least make us useless or obsolete.[7]

Each author has largely ignored the others, as if there could be many different futures.

There can only be one future. This book was written because I think that anyone hearing all the different views about our future would be wondering, as I did:

- What single future will emerge when all these different forces converge? What will happen when digital and other technologies collide with global warming? Will climate change disrupt technological advance, for example, or might technological advances help us with climate change?
- What will future rates of change be like, and how should we and our educational system prepare for them?
- Given the uncertainties, how are we to plan for ourselves and our children? What should we do now to maximize our chances of having the best possible future, given all the discordant forces?
- Will automation and intelligent robots take all our jobs? If so, what will our lives be like, how will we support ourselves, what will we do with all our free time? What will happen to us if machines subsequently reach superintelligence?
- What contingency plans should we form for ourselves and our families and what political and economic solutions should we support for our country and our global society?

This book takes a new look at these important questions. The findings to date on climate change, automation, economics, and possible digital and other technological changes have been thoroughly reviewed, and a synthesis attempted to include them in a single future. The resulting future is notably uncertain in its details, but predictable in its process, and incredible in its overall possibilities. Predictions and recommendations that have not been discussed previously come naturally from this synthesis. It seems quite possible that the future may be brighter than we think, particularly if we can manage to make opportunities out of some of the challenges we face. I hope you will consider these ideas to stimulate your thinking on your own choices and plans for the future.

This book argues that *if* we give adequate responses as a society to the problems facing us, and given the most likely interactions between all the different forces acting on us, most of the problems facing us now will be well on their way to being solved by the year 2050. A major contribution to the solution of these problems is expected to come from the rapid development of digital and other technology, including the development of effective general computer intelligence about 2035. Unfortunately, this development will almost inevitably lead to the emergence of machine superintelligence, which will present new challenges, opportunities and dangers for the human species.

The book is divided as follows.

- Chapter 2 demonstrates that, while it may be possible in some cases to predict that certain technological advances will occur, it is impossible to foresee in detail the social and economic *implications* of emerging technology.
- Chapter 3 takes a look back in time to discover what may nonetheless be predictable about the *process* of technological advance, including the tempo and growth rate of successive stages.
- Chapter 4 explores the present and near future of the digital era.
- Chapter 5 considers the probable next stage in technological development, when either digital evolution moves up to the next level or another technological breakthrough emerges.
- Chapter 6 presents evidence that normal measures of economic growth fail to provide a good estimate of our standard of living, particularly in the digital age and beyond.
- Chapter 7 investigates the problem of global warming, what is known and what not, and what solutions will be useful (and not) for the near future.
- Chapter 8 explores the future needs for education and learning for ourselves and our children.
- Chapter 9 looks at what will happen in the near future when automation and robots take our jobs, what we will do for money and what we will do with our time.
- Chapter 10 introduces the eventual challenge that we will face if machines are allowed to become superintelligent, as is the most likely outcome of our current trajectory.
- Chapter 11 provides an integrated view of our future, with some suggestions for action, as well as for some actions to be avoided.
- Appendix B summarizes detailed results and sources of data.

Sources and Further Reading. References to books are listed in the text and in the End Notes by Author (Date) to avoid writing out citations in full each time. Complete citations for all books are given in the General Bibliography. Other sources are referenced in full in the End Notes. End notes are listed in order throughout the book, to avoid having to figure out which chapter you are in (a completely unnecessary step in my opinion). In general, when it was felt useful, books (and magazine articles) were also included as sources with the thought that such discussions are often more useful to the reader than the original professional articles.

2. OUR UNKNOWABLE FUTURE

> It is impossible to predict the emergence of unforeseen technology and even the full implications of technology that has already been discovered. The point is illustrated by the example of the development of the mobile phone and its effects on world social and economic development, particularly in undeveloped areas.

There are many opportunities and potential pitfalls lurking in the list of topics heading the previous chapter, lurking behind catchy titles like digital evolution, nanotechnology, genetic manipulation, human life extension, global warming, automation, and intelligent machines. All of these topics involve opportunities that must be seized, and potential problems that must be solved or avoided.

A largely unrecognized difficulty with even the straightforward solutions suggested for emerging problems is that we simply cannot predict what human society will be like even a few decades from now, much less a century. We do not know what technologies will become available that are presently unforeseen. Even when we think we can glimpse the technology coming, we cannot foresee the implications of its development, and what the consequences will be for our society and economy. I hope to demonstrate that not only do we not know, it is *impossible* at present to predict the future in any detail even a few decades ahead. As is seen in the next chapter, however, it *is* possible to detect patterns in the changes that have been occurring, project these patterns into the future, and to use these patterns as a guide for choosing actions to take now and in the immediate future.

It will be seen that even after a technology has been introduced, the detailed implications of the technology, its effect on the society and economy, cannot reliably be predicted. These outcomes are essentially unknowable. A case example from recent history is offered as evidence for this conclusion. This may be only one example, but actually only one example is needed to prove the point.

Case example: Telephone service in Nigeria.

In the year 2000, there were about 450,000 working telephone lines in Nigeria, mostly provided by the state-owned telecom NITEL. With a population of 124 million, this meant that only four people in a thousand had a phone line. The rate of addition of new lines was only keeping up with population growth, the cost of installing a line (not including bribes) was $1,300, there were millions on the waiting list, and the typical wait was several years.[8] The majority of Nigerians, relatively poor people, simply could not hope for a phone. If you had projected phone use twelve years into the future, using past and current performance, you would have predicted that, in 2012, of a projected population of 162 million, a half million people would have telephone service - only 0.3 percent of the population.

Instead, by the end of 2012, there were 110 *million* phones in active use in Nigeria, covering 72 percent of the population.[9] Your prediction would have been low by the amount of 109.5 million, or by a *factor* of 220. As you probably guessed, the new phones are almost all mobile (cell) phones: the number of functioning fixed lines has actually declined. Mobile phones have become essential for ordinary Nigerians in daily life and business.[10] Everyone uses them for their essential contact with distant friends and relatives, their banking, taking and sharing photos, internet access, etc.

Small business owners use their mobile phones to take payments, conduct necessary banking, order supplies, send email to customers, etc. They literally could not do a profitable business without them. Last year, the growth in mobile phones and the internet connectivity they enabled were responsible for *one-third* of the increase in Nigeria's economic output.[11] There is no wait to get a phone now, and the price is, obviously, affordable for even the poor majority. This completely unexpected change in Nigerian society took only twelve years. This is not an isolated example: the rates of adoption and changes due to some other recent digital innovations are discussed later.

In the U.S., the first consumer mobile phone network was established only in 1983 (e.g., see Ryan, 2010). No one at that time foresaw the mobile phone takeover of the world, least of all how important it would become to the daily life and economy of undeveloped regions around the world. I certainly didn't. No one foresaw how many different functions could be usefully combined in a mobile phone. The first text messages were not sent until ten years later, in 1993. It was not until November, 2000, that a phone with a built-in digital camera became available to the public.[12] Routine internet access and information search using the mobile phone became common only with the availability of smart phones, such as the Apple iPhone released in April 2007 -- nearly 25 years after the first mobile phone networks.

The mobile phone is arguably the best, but not the only, example of the unexpected effects of technology on society in the present age. Nor is the rapidity of change limited to this example; see Ryan (2010) for rates of change for digital technology in general and the internet in particular. Unfortunately, peoples' thinking, their expectations for the future, and even experts' forecasts of future changes are made based on logical projection of current trends, incorporating current and foreseeable technology. Then a technology is introduced which is often completely unforeseen, and almost always with applications that no one, not even the inventors and early adapters, envisioned at the time. The time scale for the adoption of the new technology, from the time of its introduction to the time where it affects the majority of the world, can be breathtakingly fast. The old technology, supplanted by the new, suffers "creative destruction" and simply fades away.

Even in the U.S.A, where landlines are relatively cheap and easy to obtain, many younger people have never had any phone but a mobile phone, and see no reason to do otherwise. There *is* no reason to do otherwise, for most people, now that mobile networks have become so widely available.

Clearly, the mobile phone has made such fast progress because it is a very useful application, a "killer app." Its progress is the more remarkable because it requires a large investment in infrastructure before its use is routinely practical. The cost and size of the phones in the beginning also seemed likely to limit their applicability. No reasonable person would have expected them to be to be adopted so quickly by the vast majority of people on the planet, nor found to have so many uses.

While the specific implications of technological changes are therefore often unknowable, it is possible to discern some general patterns to the overall process of technological development over human history, and so provide some guidance toward the *process* expected with technological changes to come.

3. A NEW LOOK AT PAST, PRESENT AND FUTURE

> A review of economic and technological development over the past two thousand years suggests that human civilization has gone through a number of successive stages: agricultural, industrial and digital; and that the technology of each succeeding stage has overlain that of previous stages in a multiplicative manner. Consequently, growth rates increase in each stage, and each stage progresses more quickly. The *process* of technological change is therefore to some extent predictable for the near future.

Predicting the Future by Linear Extrapolation

It is first of all useful to consider the past because it makes clear the impossibility of knowing the future in detail, of understanding the implications of what is coming. This is most easily illustrated by imagining oneself in one of the past ages and being able only to understand the future in terms of extrapolating the trends of the current age, making what would be called a "linear" extrapolation. From the viewpoint of the agricultural age, for example, one would predict better carriages and roads, stronger horses and better plows, but not steam power, tractors, automobiles and trains. Further, in hindsight, it is clear that even after a new technology has appeared, it is impossible to predict the implications of the technology for society.

In the 1950s, for example, I would have predicted that by 2000 that there would be people living in colonies on the moon and Mars, as my favorite science fiction writers did. People would travel to the airport by jetpack or flying car, then take a rocket plane traveling between cities by ballistic trajectory. The only computers in people's lives would be the large ones helping to pilot rockets. No science fiction writers foresaw that everyone would have a desktop computer connected by an internet, or would have mobile phones with multiple functions that worked wherever you were. Ray Kurzweil *did* predict many features of the internet in his 1990 book, *The Age of Intelligent Machines*. However, this view was not widely shared. When President Bill Clinton convened a futurist conference in 1992, for example, *the internet was not mentioned.*[13] Other examples of forecasts especially by experts that completely missed emerging technologies are summarized by Reese (2013) and Ridley (2010).

While it is therefore impossible to predict the specific technology that will impact us even in the near future, it is possible to predict that it *will* occur, and in many cases there will appear new technology, or at least applications, that are *unexpected.* More important, we can get some ideas about what will come by looking at the *pattern* of the stages in the past, as well as the accelerating pace of change for each succeeding age. While the content of the future is quite uncertain, in other words, the process may be slightly more predictable at least when understood from a new perspective.

A New Overview of Our Recent Past and Near Future

To gain this perspective, we must take a brief overview of our recent history, the development of human civilization and the associated rates of change through the past few thousand years. It will be seen that human culture has gone through several clearly delineated stages, and that there has been an amazing and rapid increase in the *rate* of change typical of each successive age, and a corresponding rapid decrease in the time each stage lasted (or is predicted to last).

To illustrate changes over the ages, we consider first two traditional measures, world population and average income per person. These two measures are then compared with measures of technological progress, including computation speeds, numbers of cell phones sold worldwide, numbers of industrial robots installed, and other measures. For those who wish to see the detailed results, sources for all data and precise numerical results are presented in Appendix B. The main results are summarized in Figure 1 and in the text below. **Note:** For our purposes, the average annual income per person is estimated by dividing the estimated gross world product (GWP) by the total population. Cautions and limitations in using the GWP for this purpose are discussed further in Chapter 6.

During what is called the agricultural age, lasting 8,000 to 10,000 years, most of mankind lived in rural areas or small cities and were involved with food production. The total population, limited by food supply, grew very slowly, and the personal income grew even slower. Human power during the agricultural period was gradually supplemented by animal power, and food production increased through agricultural technological improvements, such as the steel plow, crop rotations, etc.

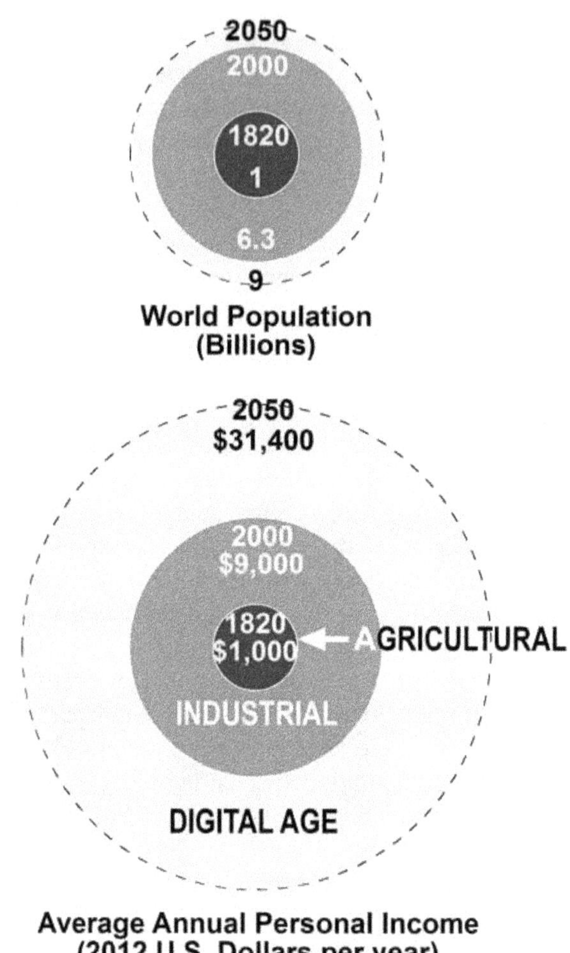

Figure 1. Overview of Human Development to 2050 CE
(Using conservative projection for future income)

By 1820 CE (current era) the world population had reached about 1 billion and average income was $1,000 per year. (See Figure 1; areas in this graph are proportional to magnitudes, and monetary results are expressed in constant 2012 U.S. dollars.) During the industrial age, steam power and later fossil fuel power increasingly were used to supplement human physical power, and machines augmented human physical performance. Total population grew very fast, and personal income grew even faster: by 2000 there were 6.3 billion individuals earning $9,000 per year on average. Population growth rate started to decline as people moved from rural to urban settings.

By 2000, the digital age began, when human intellectual and physical manipulation began to be significantly enhanced by digital processing (discussed below and in Chapter 4). While the population growth has slowed, predicted to peak at about 9 billion about 2050, the average income per person has continued to grow exponentially. The conservative projection shown in Figure B3 and Figure 1 illustrates the case where the personal income continues to grow at the present rate of about 2.5% per year until 2050. At this point, personal income would reach $31,400 per year.

Characteristics of Cultural/Technological Ages

As indicated above, it seems useful to categorize human technological and cultural changes by division into several characteristic "ages": so far, the hunter/gathering, agricultural, industrial and digital ages have been identified or labeled. Before proceeding further, however, it will be useful to more clearly define what the implications of such divisions might be and how different ages and their characteristic technologies might relate to one another. This definition will be necessary to understand the motivation behind the identification of different ages and the relationships of the technologies and growth rates associated with each age.

Consider first the processes which characterize the agricultural age. Humans, starting in a hunting and gathering stage, began farming crops about 12,000 years ago. At first, farming was just a supplement to the "gathering" part of the society. As farming became more successful, gathering decreased further and even hunting became supplemental to farming activities for most societies. At this point, one could say that an agricultural age had begun.

An industrial age was *added onto* the agricultural age in the early 1800s when factories began using the steam engine and machines to facilitate production of consumer goods. The precise date for the start of the industrial age varies from one author to another; for our purposes it was determined primarily by noting the sharp increase in personal income that occurred about 1820 (Figure B1, top curve). At that time, industrial output exploded, the iconic product being ready-made clothing that the average person could afford. Later advances in this age were characterized by development of trains and automobiles.

The agricultural age did not *stop* in 1820, of course, rather agricultural processes were multiplied by the overlay of industrial technology. Tractors running on fossil fuels pulled steel plows, fossil fuel fertilizers were introduced and machinery helped in the harvest, all made available inexpensively by the society's industrial output. While the factories of the industrial age were mainly located in a few regions, the effect of the industrial age was felt over the whole planet, as evidenced by the fact that the world population exploded.

These two historical examples suggest that the definition of a global cultural/technological age might carry the following implications.

1) The technologies or innovations characteristic of the age have affected society significantly, to the extent that they have become central to the lives of most people on the planet.
2) The technology of the new age has overlain earlier technologies in a *multiplicative* fashion, such that important and widespread advances are produced that could not be obtained from earlier technologies alone.
3) The new technology allows at least some parts of society to leapfrog change, jumping over steps that would have been required if they had been restricted to the previous technology.

Using these criteria, it is argued that another new age has recently begun, the *digital age*. As a prime example, consider the use of mobile phones. The number sold on the planet began to rise sharply about 1997, until last year the cumulative total sold equaled the world population. Currently, the mobile phone is not only being used by nearly everyone on the planet, it is being used in ways that have significantly affected the average person's life (Chapter 2). The mobile phone is a (multiplicative) combination of the industrial age telephone and digital technology. It also has allowed societies to leapfrog change, as many people now have a mobile phone who have never had a landline phone, and many still don't even have electricity.

The start of the digital age is here proposed to be taken (somewhat arbitrarily) as the year 2000, on the basis of the growth of digital technology and its use by the majority of people on the planet. The evidence for this transition is summarized in Appendix B. Note that, just as the industrial age "overlaid" the agricultural age, but did not *replace* it, so the digital age is posited to overlay and contribute to the technologies of all previous ages. This concept has important implications for typical growth rates of successive ages. It will be seen (and/or hypothesized) that this overlaying process has caused (will cause) growth rates of personal income to *increase sharply as a result of each transition.*

Growth Rates in Different Ages

The most relevant growth rates, representing the *rate of change* (slope of lines in Figures B1-B3) are summarized in Table 1. (Growth rates and doubling times for exponential growth are discussed in Appendix A.)

Table 1. World Population & Personal Income Growth Rates

Age	Over Years	Population Growth Rate	GWP/Person Growth Rate	GWP/Person Doubling Time
AGRICULTURAL	1000 - 1820	0.15%	0.06%	1,150 years
INDUSTRIAL	1820 - 2000	1.0%	1.3%	54 years
DIGITAL	2000- 2050?	declining	2.5% min.	28 years max.

Growth rates during digital age are minimum projections: See text.

For the agricultural age, values given in Table 1 are restricted to the most recent period 1000-1820 CE, for which data are more reliable. The growth rates projected for the digital age are discussed in the next chapter, while the rates observed for the first two eras, and other characteristics, are discussed next.

The era of agricultural innovation. This era extends from the establishment of agricultural societies around the world until the start of the industrial revolution in 1820. The agricultural era was characterized by a very slow but steady rate of invention and implementation of agricultural technology, which allowed the slow growth of world population, limited primarily by food availability. Key innovations included the iron plow, the use of draft animals, development of new crop species and rotation of crops, among others. These innovations allowed a slow increase in world population size from an estimated 240 million at the start of the first millennium to about 1 billion by 1820. The average annual population growth rate over the period 1000-1820 CE was only 0.15 percent. On average, the population doubled every 460 years over this period.

The slow development of agricultural technology during this era not only allowed the population to increase, but more importantly, it allowed the food productivity per person to increase (albeit very slowly). From 1000 to 1820 CE, the average income increased from $624 to $1,040 per year (expressed in constant 2012 dollars). This is a very slow growth rate, an average of only 0.06 percent per year, or a doubling time for GWP/person of 1,150 years. During this era the average person saw essentially no change during their lifetime.

There were, of course, also important non-agricultural innovations during this period, such as the printing press and an increase in world-wide trade. Because slightly fewer people were required to grow the same amount of food and because of the rise of other trades and of trading, there was a corresponding slow growth of non-farm labor during this period. By 1820, for example, 12 percent of the population in Japan and Western Europe lived in cities of over 10,000 (Maddison, 2006). The availability of this non-farm labor initially facilitated the much more rapid changes that accompanied the industrial revolution.

The industrial revolution. Changes in society during the period 1820-2000 were initially driven by the invention of the steam engine and steam-driven devices that increased the average productivity per worker. The increased production mostly occurred in a relatively limited part of the world: England and Western Europe, North America and Australia. However, the change was massive, and affected the whole planet. The total world population increased from about 1 billion in 1820 to 6.3 billion by the year 2000, an annual rate of 1 percent. Even in the face of the rapidly increasing population, the GWP/person also increased greatly, going from $667 to $6,756 over the same time period, an *increase by a factor of ten* over only 180 years. This corresponds to an annual growth in personal income of 1.3 percent, equivalent to a doubling time of 54 years. This is *22 times* the average growth rate characteristic of the agricultural era just prior to it. In contrast to the case in the agricultural age, the average person in this society would clearly have experienced significant change over a lifetime - provided his or her lifetime was a bit longer than average, that is.

4. THE DIGITAL AGE

> The digital age began with the explosive growth of computer technology, which began having a significant impact on global society about 2000, affecting all areas of human communication and knowledge acquisition, industrial production and distribution, and transportation. Near future developments include growth of computer intelligence and digital contributions to improvements in living standards.

Growth Rates of the Driving Technology

The stage for the even more rapid increases which characterize the present era was set by the invention of the transistor, and the combining of transistors into a single integrated circuit ("chip"). Since then, all of the important characteristics of computing have continued to increase exponentially, with a doubling time less than two years. It is suggested here that the digital age effectively began about the year 2000, when the increase in processing speed enabled applications which began to have a measurable effect on the average person on the planet. (See Figures A3, B2, B3 and 6, and Table B1)

Current rates of increase of different measures of computer speeds and memory are summarized in Table B1. The only difference between the different measures is a small variation in doubling time, *all of which are amazingly short - only 1.0 to 1.5 years.*

Will the exponential growth in computer characteristics continue? From time to time, a computer expert claims that computer speeds or memory have reached a natural limit, or will do so soon. All we know for sure is that, in defiance of such predictions, exponential growth has continued for many decades. As shown in Figure A3, computation speed per $1000 has grown faster than exponential for over 110 years (including computation devices used before electronic digital computers). In the absence of any actual evidence to the contrary, we will therefore assume that growth will continue at the current rates. At the least, this results in concrete predictions, while the alternative case does not.

Impact on Society of Computer Power Increases

Some experts doubt that the predicted, continued exponential growth of integrated circuit power will have a corresponding impact on society, see, e.g., Meisel (2013). His argument is that the speed of many applications (e.g., word processing) depends on user input and increases in computing power will not affect the productivity of that process. However, it has *not* been shown (and will be argued otherwise) that *no* important process in society will be affected by the relentless exponential increase in digital processing speed. That is, it is argued below that the steady doubling of computer speed will continue to enable applications which are potential game changers and (often) completely unexpected.

We have already discussed one example showing an exponential growth in application: the worldwide rise in mobile phone use. The total number of mobile phones sold worldwide is shown by the solid circles in Figures B2 and B3. The rise was clearly exponential, with a time scale on order of two years, with the most recent numbers probably slowing only due to saturation of the market. Nearly everyone in the world now has a mobile phone, and this change took place over only 15 years.

Note that the drive in mobile phone usage must have been enabled (at least partly) by the increase in computation power per dollar: The decrease in chip price (for a given number of transistors) along with economies of scale allowed the price of mobile phones and mobile networks to decrease rapidly so that more people could afford them.

Typical reductions in cost are illustrated by the example of Nigeria discussed above. When mobile phones were first made available in 2001, the total cost to get online was $600, an amount met with outrage in a country where this would have been 6 months wages for the average worker.[14] By the present day, rates have fallen to the extent that over two-thirds of the people in the country have been able to afford a mobile phone (though not without some sacrifice by the poorest).

Since 2000, the annual rate of increase of the average global personal income has been about 2.4 percent.[15] Part of the reason there was a respectable increase, in spite of some economic problems, is that the total world population has not been increasing as fast during the last decade as it had previously, so that increases in productivity were not split among a rapidly growing population, as they were during the industrial revolution. The GWP per person per year is now about $13,000 (2012 dollars).

At least some fraction of the increase in personal income in the present era can be directly tied to the rise of digital technology. In Nigeria, as mentioned above, information and communications technology (ICT) alone was responsible for one-third of the increase in the country's gross domestic product last year. Ridley (2010) suggests that this pattern is quite general for developing economies in Africa and elsewhere.

Digital Age is a New Level of Social Development

It is suggested that the digital age indeed qualifies as a new stage in the development of human culture.[16] Point by point (to compare with the general definition offered in Chapter 2):

1) Digital technology has made a significant and world-wide impact on the population of the whole planet. For example, most people on the planet have a mobile phone and are connected to the internet. These connections have significantly affected life on the planet, enabling personal and business activities that could not have existed before.

2) The modern mobile phone is a perfect example of the marriage of digital technology with several industrial age technologies: the landline telephone, the film camera, snail mail, catalogs, phone books, newspapers, libraries, etc., to yield a product and social activities that could not exist without ideas and technologies from both eras.

3) Being connected by mobile phone, for example, has allowed people in undeveloped countries to leapfrog past the need for fixed lines. Many in Africa, for example, have mobile phones but not yet electricity.

As Table 1 shows, the first three eras are characterized by a rapid stepwise increase in the speed of advance. The agricultural era took about 8,000 years; the industrial era only 180 years, and the first part of the digital evolution, the digital era, is expected to last only 35-50 years, as discussed next chapter. There was a steady increase in the rate of growth of personal income, from near zero at the start of the first millennium to a rate between 1 and 4 percent per year over the last 200 years. The correlation of the decrease in the length of the age with the doubling time is suggestive, and considered further in the next chapter,

The difference between the two most recent eras, industrial and digital, is that the main gains during the industrial revolution were due to improvements in machinery which allowed the physical productivity of the average worker to increase rapidly. The productivity in this period was primarily in goods and services; the iconic products of the era were manufactured ready-to-wear clothing, and later the automobile. The digital age, in contrast, is marked primarily by the supplementation of human intellectual capacity, with a huge increase in communications, information and entertainment, the iconic products being desktop computers, the internet, mobile phones, and automation.

The digital age is distinguished from all earlier eras by the incredible, exponential growth rate of its driving technology and applications (Table B1). Such exponential growth was not characteristic of the forces driving the industrial or agricultural ages. The change rate was so slow in the agricultural age that the average person would not have noticed change during their lifetime. Changes did speed up in the industrial age. There were stepwise improvements in factory technology, engine technology, energy production, etc., and gradual declines in costs but nothing important changed exponentially over any length of time, certainly not with a time scale of only a year or two.

Over many decades, in contrast, computer speeds have been increasing exponentially, with a doubling time less than two years. Another way to look at the same statistic is that the cost of a given amount of computation power has been dividing in half every year or so for over a century. This has made many applications feasible that would have been limited by cost considerations, some of which are discussed below. Digital technology is expected to be a major contributor to the (conservative) predicted doubling of personal income in less than 30 years. With the average lifetime stretching toward a century, the average person in the digital age will clearly experience significant change in their lifetime. Some of the changes that have already occurred are summarized next.

Characteristics of the Digital Age So Far

The digital automobile. While the rise of digital technology is arguably best illustrated by the mobile phone (Figures B2 and B3) this is not the only relevant consequence of the combination of inexpensive digital technology with previously existing technologies. Another good example is the modern automobile, a combination of digital technology and the automobile of the industrial age. The industrial age automobiles were polluting, unreliable and had relatively poor gas mileage. Part of the solution to these problems involved the marriage of a catalytic converter with digital control of the engine. It was found that putting a catalytic converter on the engine exhaust could remove many of the worst pollutants (unfortunately not including CO_2). Also unfortunately, such a converter cannot function unless the engine is run very lean.

Running an engine lean means using the lowest gasoline/air ratio possible, and this is also the most fuel-efficient mode. This would be a win-win solution, but running lean cannot be done practically by normal human or mechanical control, because a motor set up to do this tends to stall when an attempt is made to accelerate it. Modern cars manage to run lean and to run efficiently by adjusting engine characteristics on the fly using an onboard digital controller. This increases fuel efficiency and has the added benefit of making cars much more reliable, e.g. they start every time without any particular skill on the part of the operator.

The non-digital automobile. In this respect, for those who have not had experience with earlier automobiles, let me report my experience with the 1950 Ford that I owned in high school. I cannot count the number of times that my car inexplicably wouldn't start, even though I kept the engine in good condition. I might have "pumped" the gas the wrong number of times before trying the starter, thereby "flooding" the carburetor, or the problem might be that the "automatic" choke was stuck (being a mechanical device). Every day was an adventure with this car.

Near-Future Changes in the Digital Age

The speed of recent changes in the many applications of digital technology, most clearly not foreseen, has taken everyone by surprise. Clearly, it makes the future difficult to predict and just a little bit scary. Some of the changes predicted for the near future, most of which already appear to be in progress, follow.

The software connections between people and computers will continue to improve, allowing the best use of these two different types of intelligence. For example, Meisel (2013) notes that error rates for speech recognition software are decreasing at 18 percent per year. Applications will increase rapidly as computers are able to better understand and accurately act on human commands. For example, Google smart phone search is now moving (has moved) toward a model in which their search engine answers the question asked (as Apple's Siri already attempts to do) rather than giving a list of web pages which are related to the query.

Computer-driven automobiles. Nearly all automobiles sold in the developing world already have a large number of digital processors installed. The engine control system, discussed above, is already standard, as are (or are becoming) automatic braking systems (ABS), stability control, skid control, emergency radar braking, adaptive cruise control, and services providing directions based on the global positioning system (GPS). Google among others is now developing and test-driving completely robotic driving systems. Robot driven vehicles autocars, autotaxis, autobuses could provide many ecological benefits while enhancing the quality of life.[17]

Wearable computer, information and virtual reality systems. In 2006, Verner Vinge produced a science fiction novel, *Rainbows End*, which arguably contains the best description of the kind of technological changes that may occur in the near future, involving wearable, voice-controlled computers connected to the cloud, giving access to instant information and virtual reality nets, plus distributed nets and computer driven automobiles. All of these predicted devices are already on the market or in development, e.g., Google glass, 3D augmented reality goggles,[18] smart nets and the "World of Things."[19]

Transportation systems of the near future. Combining all these digital advances suggests the following vision for a transportation system of the near future. Suppose you decide that you want to go somewhere too far to walk. You would simply say, "Central Dispatch, please send me my car." This communication is, of course, by voice command through the equivalent of a smart phone that you are wearing, and Central (a supercomputer) acknowledges by saying into your ear, "Car will be at your location in 3 minutes." Since

Central knows where you are (it always knows who and where you are, of course, due to the GPS enabled devices you are wearing) and it knows where it parked your car, it sends it to you. If you don't have a car, or don't want to use it, Central finds the nearest unused autotaxi in an underground parking/charging facility, and brings it to you.

When you get in, you tell the car where you want to go. Central calculates the fastest route, knowing the traffic (Central always knows all the traffic) and puts the route on the dash screen, asking you to approve it. You decide it's mostly fine, except you indicate, by touching the screen, that you want to drive on the street past the park. It accepts your change and drives you to your destination, while you relax and read the day's news on your heads-up display. Or listen to music on request. Or work online.

When you get to your destination, you simply get out and walk into the building. The car parks itself back in an underground parking/charging station. The car is, of course, an all-electric car, run by batteries charged up at stations ideally using electricity from solar or other non-polluting power. If you have to go farther than you can travel on a single charge, Central can simply have another car meet you where needed. Alternately, especially if using your own car, you could stop at one of the many drive through businesses who will give your car a fast recharge while offering *you* a recharge of coffee and muffins. Or a full dinner, manicure, etc.

Population peak. One of the most obvious characteristics of the digital age noted in Figure 1 has no obvious connection to the digital revolution: the population of the whole world is projected to stop increasing for the first time in history, reaching a maximum of about 9 billion around 2050. This phenomenon is actually an end consequence of the industrial revolution. The urbanization of the world has caused an effect known as the "Demographic Transition": the rate of population increase is observed to drop when *any* civilization goes from primarily rural to urban. The reasons this happens apparently depend on many factors including the education of girls and the jobs available in urban civilizations. Underlying all possible factors, however, the demographic transition is due to the fact that women stop having as many babies. After all, children on a farm are extra hands, but children in the city are just extra mouths. And newly educated women can profitably work outside the home in a city.

Expected Growth Rates in Emerging Digital Age

The prediction of a peak in population, or at least a slowing of population growth, seems relatively firm as it already has started. The growth rate of personal income that one might predict for the digital age ahead is necessarily much more speculative. *The digital age is only a little more than ten years old,* so it seems unlikely that the current growth rate is a valid predictor of its future potential. Consider, for example, the faster than exponential growth that occurred as the industrial Revolution took off (Figure B2). In any event, the growth rate of GWP/person since 2000 has been about 2.4 percent.

Making the relatively conservative assumption of no new breakthrough technologies in the next forty years, the result of Randers' (2012) dynamic systems modeling was a relatively steady 2 percent annual growth rate in GWP/person over the entire period. However, it seems highly unlikely that there will be *no* additional acceleration effect due to improved technology. General Electric predicted in a November 2012 report that the "industrial internet" alone has the potential to add $10-15 trillion to GWP by 2030. This would imply an increase in GWP from industrial internet applications alone of 1 percent per year, giving, for example, 3 percent growth in GWP/person instead of 2 percent. Along these lines, the current International Monetary Fund prediction is for a 3.2 percent increase in GWP/person in 2014.

Given all this, a 2.5 percent annual growth rate in GWP/person seems a reasonable, quite conservative estimate for the immediate future; this estimate is the one shown in Figure B3 and Table 1, leading to the results summarized in Figure 1. In a later section (Chapter 6) it is shown that GWP significantly underestimates real growth, especially for the digital age, because it ignores free services and quality improvements in goods, services, and overall living. Therefore, I will argue later that the "Real GWP" is significantly higher than the traditional estimate and will increase much more rapidly.

In addition, as digital technology takes over an increased share of the GWP in the immediate future, it is reasonable to expect that even the traditional GWP will begin to rise faster simply because it will be pulled by the very fast exponential rise in digital technology. Therefore, in a subsequent chapter an overall growth rate of 4 percent or more in "Real GWP" is predicted for the rest of the digital age, compared to the 2.5% rate used in Figure 1.

What Impact Will the Digital Age Have?

Some mainstream economists think the major innovations are behind us. Robert J. Gordon is quoted as saying, "the real impact of the electronic revolution has already occurred."[20] On the contrary, most experts think that most of its impact is ahead of us. Just taken by itself, full automation will have a huge impact on global society and is just beginning. Up to now, companies in developed countries have been able to put off full automation by outsourcing jobs to undeveloped areas where wages are still low. As these economies feel the benefit of these jobs, wages there rise, and as costs of industrial robots continue to fall, there will eventually be a drive to full automation for industries all over the planet. The changes in society from this effect alone are expected to be immense, discussed in Chapter 9.

In addition, the use of robots by the military is just beginning; full armies, air forces and navies of semi-autonomous robots are in the future. Home, school and medical robot sales have just begun to take off; completely automated housework, education and medical care are coming. The numbers suggest the process is just starting: worldwide, in 2011, there were 6,500 military robots sold, of which over 5,000 were unmanned aerial vehicles (UAVs). There were 16,000 professional service robots sold, of which over 1,000 were medical robots, mostly surgical. Other service robots sold included rescue and security robots, construction and demolition robots, underwater systems and milking machines. In addition, a total of 2.4 million household and personal robots also were sold at an average price of $250 each. These included toys and robot kits but also lawn mowers, vacuum cleaners and other household devices.[21] And automation and robotics represent just the part of the ice berg that we can see, there are all the completely unexpected technologies and applications to come that will unfold just as the mobile phone, digital photography and the internet did.

Other effects of computer speed increases. A number of useful applications will become possible *only* with an increase in computing speed. The doubling and redoubling of computer speed is relevant to the extent that it enables applications which go on to have a significant impact on society. For example, ten years ago the computer time to generate a typical model for 3-D printing could be several weeks.[22] At present speeds, it takes several hours, which in fact makes it possible for this technology to have a much larger impact.

In many areas of scientific investigation, the speed of computer calculation available currently is the main limitation. In modeling of physical processes such as fluid flow, for example, the accuracy of model simulation increases as the model is made more fine-grained, so that more data points are analyzed in each time step, which also needs to be reduced to improve accuracy. As a particularly relevant example, models used to predict future climate changes due to global warming have clearly been of limited value at least partly due to limitations imposed by current supercomputer speeds.

Further, expected increases in computer power, or equivalently, having more computation available at given size and power requirements, makes *any* application more available: available in different configurations, available at a given cost, available as part of a cloud or mesh system, etc. Appendix C, for example, presents concrete predictions for the future power of wearable and implantable devices. The combination of increased computation power in smaller units using less power[23] with recent improvements in battery technology and wireless communication will also allow many distributed systems to be developed which can only be conceptualized now. One example would be the development and distribution of a large number - a worldwide mesh - of weather sensors to improve the science of weather and climate forecasting. Of course, it remains to be seen the extent to which *any* of these new applications of computers actually causes a significant change in human society. ***But the chance that none of them will seems vanishingly small!***

All of this suggests that the digital age has just begun. Compare the digital age with the industrial age preceding it (Figures B1 and B2), especially the growth rates in GWP/person just after the start of each age. Fifty years into the industrial age, in 1870, the growth rate had only begun to really take off. We are, according to our hypothesis, only 13 years into the digital age. Even accounting for the time compression of modern changes, it seems likely that our current age has just begun. We agree with John Pierce when he says, "*After growing wildly for years, the field of computing appears to be reaching its infancy.*"[24]

5. DIGITAL PLUS: THE NEXT STAGE

> The digital age, though just begun, is not expected to last long before technology is enhanced further by the next major technological gamechanger. This could even be digital technology itself, which may bootstrap itself into a new level when intelligent machines become significantly smarter than humans. This process should begin to have a measurable impact on society about 2035, marking the end of the first digital age. Alternately, other technologies may emerge by that time to boost our culture to a new level by overlaying earlier digital, industrial and agricultural technology.

The process will not stop here. Just as industrial technology pushed agricultural technology into the background, and began to multiplicatively combine with it, and as digital technology has begun to multiplicatively combine with industrial and agricultural technology, we can expect another stage to take off soon, as a new technology multiplicatively begins to combine with digital technology and the layers below.

However, we cannot accurately predict now what technology that will be. It could be a leap forward due to a bootstrapping or regenerative effect of digital technology, as discussed below; or it could be the addition of one of several technologies now in early stages, as proposed by Gore (2013), Mulhall (2002) and others; or it could even be a technology or application currently unrecognized by anyone. After all, even as late as 1992, few expected the internet takeoff, and none predicted the worldwide impact of mobile phones.

If Digital Technology Bootstraps Itself Into a New Stage

Digital technology could bootstrap itself into a new accelerated phase if computer intelligence continues to increase exponentially. There are two stages that can be distinguished. The first stage would be when computer intelligence first *generally* exceeds human intelligence, marked by the achievement of artificial general intelligence (AGI). The second stage would occur if and when computers became superintelligent, that is, able to function autonomously *and* to design computers who are smarter than they are. This stage, also known as Artificial Superintelligence (ASI), would result in an exponential explosion in machine intelligence, probably leaving humans well behind. Discussion of this second stage is deferred to Chapter 10.

Current expectations are that computer intelligence will generally "exceed" human intelligence by about 2025, and it is estimated below that computer intelligence will be significantly greater than human intelligence by 2035. That is, it is expected that autonomous, intelligent machines will begin to make important impacts on human society by that time. Most experts expect that this innovation will cause an additional quantum leap in technological and social change at that time.[25] In other words, whenever it occurs, the achievement of AGI is expected to create a change by itself yet again as large as the advent of the (first) digital age has done.

Specific predictions for the time scale for development of artificial general intelligence (AGI) have been made by Kurzweil (1990, 1999, 2012). He has accepted a $10,000 bet that a computer will pass the "Turing Test" by 2029; that is, it will programmed well enough that it will not be possible from the content of its "chat" responses alone to know that it isn't human.[26] He states, "My own consistent prediction is that this will first take place in 2029 and become routine in the 2030s." Obviously, a significant milestone along this path was recently reached by a huge computer named Watson, which won a game of Jeopardy against two top human players (see analysis by Kurzweil, 2012). However, Watson could not have passed as human; some of his mistakes would have exposed him (and note the pronoun shift).

As mentioned above, Turing suggested one test of computer intelligence which has been given his name. Contrary to the way this test is usually understood, however, Turing actually suggested it to *make the opposite point* (Cowen, 2013): Neither the ability of a computer (or anyone) to pass this test *nor* the failure to do so is a good judge of machine (or human) intelligence! There are many machines today who are much more intelligent than humans in one or more areas of application: playing chess, Jeopardy, analyzing genomes, stock prices, etc., and none of these machines can pass a Turing Test. There are intelligent humans, as well, who couldn't (e.g., autistic individuals). Because today many computers are more intelligent than humans in *some* area, computer experts now define the next stage to be artificial *general* intelligence (AGI), defined as a computer more intelligent than humans in (nearly) all areas - and certainly one who could pass the Turing Test *if it wanted to*, or was programmed to do so.

Simply passing a Turing Test in any case is not sufficient for AGI since it merely means that a computer could be programmed to *appear* as intelligent as a human, not that its intelligence would necessarily be of practical, significant value to humanity. After all, there are plenty of humans who appear to be as intelligent as other humans, and humanity has not necessarily benefited tremendously from any particular one of them. One has to expect that, even when a computer does become more intelligent than a human in the general sense, it will take a few more years to develop computers that are sufficiently *enough* smarter that they could potentially impact the society, and *then* some time for these applications to mature enough to have an actual, significant impact on our planetary society.

Human brain emulation. Another way to approach an estimate of the time scale relevant to machine intelligence is to perform the following "thought experiment:" to determine at what point a computer would be fast enough to emulate a human brain, that is, to imitate our brain functions in detail, neuron by neuron, in real time? Since we do believe that we do have a kind of intelligence, one way that we could be sure to produce a computer that was actually intelligent would be to imitate human brain function down to the cellular level. This will be possible because electrical functions of neurons themselves have been well characterized mathematically for some time, and we are now discovering details of the important chemical signaling processes and of the many specific connections between neurons in the human brain. The primary benefit of such a model is expected to be understanding human mental processes like consciousness and different mental abilities, disabilities and illnesses[27] - *not* as a practical approach to AGI (see below).

What kind of computer speed would human brain emulation require? Consider that the human brain has about 85 billion neurons (8.5×10^{10}), and about 1,000 to 10,000 synapses per neuron, for a total of about 10^{14} to 10^{15} synapses. So far, the best brain model that has actually run is a model of the human visual cortex, having about 10^{13} synapses, about the same size as the cat cortex, and 10-100 times smaller than needed for a full human brain emulation. This model, using simplified neurons, ran on an IBM Blue Gene /P computer having a speed of about 5×10^{14} FLOPS, and achieved a model speed equal to 1/100 of real time.[28] This means that a human full-brain emulation achieving real time modeling speed would require a computer about two factors 100x100 = 10^4 faster, or a speed of 5 x 10^{18} FLOPS. The appropriate measure of computer speed to use for this application is the speed of the world's fastest computer (Table B1). The world's fastest computer presently has a speed of 3 x 10^{16} FLOPS, and the speed of these computers has been doubling every year. If this continues, the speed of the fastest computer will hit 10^{19} FLOPS about the year 2020, approximately the speed needed for this level of brain emulation.

Developing a model that includes much more detail about individual neurons, Markram has suggested that by 2023 computer speeds and memory will be sufficient to acurately model the human brain.[29] Again, this is an important step, but only a step. To actually attain human intelligence such a brain model would have to trained, to somehow learn about the world from a tutor, as a human infant learns from interacting with an adult human and its environment. It will take some time to figure out how to best do this, and to give the simulated mind the necessary experiences and knowledge. Kurzweil (2012) discusses this problem, noting there is a bootstrapping problem if the computer has to be taught by another computer. If the learning took as much time as human development does, of course, it would take 20 years for a computer to reach a mature level of knowledge.

As noted, computer emulation of a human brain is primarily important, for our purposes, as a thought experiment. It proves that there is *one* way to produce a computer intelligence that equals human intelligence, and as such, this computer intelligence could have all the attributes of a human brain, including consciousness, awareness of self and the experience of free will.[30] Rather than slavishly copying our biological brain, of course, there certainly will be developed much more efficient algorithms to implement intelligence in computers that will make much better use of a computer's strengths, e.g., as demonstrated by the programming used for Watson (Kurzweil, 2012).

Taking all the above factors into account, it seems a safe bet that computer intelligence will be significantly greater than human intelligence sometime after 2023. Certainly, it seems likely that computer intelligence will have a significant impact on human society by the time computers are many thousand times faster than they will be in 2023, i.e., by 2035. Another way to estimate the effective date is to note that it took more than 30 years for the mobile phone to grow from the first network until almost everyone on the planet had one. Even given a speedup by a factor two (discussed later) it seems likely that it would be ten to fifteen years after computer intelligence reaches human level that this would have a significant impact on global human society, i.e., by 2033-2038.

All the above estimates converge to suggest that computers will have an effective intelligence significantly greater than human intelligence by about 2035, reaching the capacity for effective artificial general intelligence (AGI). In other words, I expect by that time that computers will have functional, independent intellectual capabilities that will be making a significant impact on human society, in the sense that computer intelligence will be acting on society in a regenerative, or positive feedback manner, potentially ushering in the "Next Stage."

The above arguments suggest that, if computer speed and memory continue to improve at the current exponential rate, computer intelligence will soon considerably exceed human intelligence. As a consequence, the subsequent growth of knowledge will be largely the domain of the intelligent computers. Not everyone agrees. For example, Meisel (2013) argues that computer intelligence is fundamentally different than human intelligence, and that the main use of computers in the future will involve collaborative work between human and computer, with each contributing their unique skills. He further states that we would not want to develop a computer that passed the Turing Test because we would have to program a computer to lie to us. I agree that computer intelligence is and will remain *different* than human, however, that doesn't mean that we *couldn't* develop a computer that would pass the Turing Test or otherwise appear to have human-level intelligence. Programming a computer to be indistinguishable from human intelligence is very likely to be done as soon as it can be done, if only for demonstration purposes. Look at the popularity of the chess playing computer Deep Blue and the Jeopardy player named Watson.

Along similar lines, Reese (2013) suggests that people won't want to talk to a machine that is given human traits. Nonsense, I say, people like to talk to their *pets*. And C3PO and R2D2 show that robots can be lovable, even if they can only beep. In any case, it seems unwise to say that anything is impossible or *never* will happen regarding computer intelligence. "Never" is a very long time, especially toward the end, to paraphrase Woody Allen. For those who want to pursue this issue further, Ford (2009) offers an appendix which summarizes reasons that machines can be expected to become intelligent enough to replace many jobs held by humans. Kurzweil's books are obvious additional sources for well-informed and, so far, very accurate speculation about the digital future.

Other Emerging Technologies

Of course, it may happen that computer speeds reach a physical limit and cease increasing exponentially in the immediate future, so that the next stage is not inaugurated by intelligent computers. Given current technological advances in many fields, however, it is undeniable that *some* technology will emerge to become a dominant factor and usher in the next stage. Further, even if computer speeds do not increase as fast as we expect in the immediate future, the development of these other technologies, such as nanotechnology, quantum and biological technologies, etc., is likely to lead soon after to substantial increases in computer performance - i.e., nano or quantum computers.

A number of these different emerging technologies are discussed by Hammerly (2012) and Mulhall (2002). All of them are in early stages, and any one at some time in the future could become a game changer to the same extent as the initial development of digital technology has been. The list of major contenders starts arguably with nanotechnology, controlling the structure of machines and objects on the molecular level. The second major contender involves manipulations of living organisms through genetic engineering, while a related area involves new treatments for diseases, life extension, cognitive enhancements, etc. These issues will not be discussed further here, because, as has already been demonstrated, it is impossible to predict now which technologies and which specific applications will actually have a significant impact on human society.

Note also that, while many of the potential new technologies do *depend* on digital technology, they are not inherently digital in nature. That is, data on genomic sequences may require a computer for adequate analysis, but the sequencing and manipulation of DNA is inherently a biological and chemical technology, requiring, for example, the use of the polymerase chain reaction (PCR) to amplify DNA fragments. Similarly, manipulation on the nano scale will require new understanding of the dynamics of molecules on that scale. This will require computers for modeling, and possibly control of assembly, but the technology developed will be generally based on physics, chemistry, metallurgy and biology, not computer science. As another example, the exciting development of 3-D printing requires a computer for control, but its core functions are based on innovations in materials science and chemistry. In all these potential technologies, of course, digital technology functions as an enabler and multiplier. It is the *convergence* of digital and industrial technology with one or more new technologies which makes the future era the age where the next stage begins impossible to predict in detail. However, it is possible to predict that *the growth rate will increase sharply with each new layer of overlapping technology.*

Because the next technology to take off is still uncertain, the next stage cannot be named at this point. The appropriate name might turn out to be "Age of Intelligent Computers," or "Age of Nano Engineering," or "Age of Genetic Engineering" or something else whichever becomes the next major game changer, multiplicatively combining with previous technology. Growth rates and dates are volunteered in Table 2 on the basis of the estimates above and the pattern suggested in Table 1. Note that the growth rates for the last two ages in Table 2 are for the "Real GWP" as discussed in the next chapter, that is, a GWP or equivalent measure that takes into account the provision of free services, and the improvements in the quality of goods and services characteristic of the digital age and beyond.

Table 2. If the Pattern Continues....

Age	Years	GWP/Person Growth Rate	GWP/Person Doubling Time	Length of Age / Doubling Time
AGRICULTURAL	**up to 1820**	**0.03%**	**>2000 years**	**< 4**
INDUSTRIAL	**1820 - 2000**	**1.3%**	**54 years**	**3.3**
DIGITAL	**2000 -** 2035	4%*	18 years*	2.0
NEXT STAGE	2035 to ?	8%*	9 years*	?

* Refers to "Real GWP": See text.

The rightmost column in Table 2 is provided for further elaboration of a suspected pattern. This column presents a dimensionless ratio, equal to the length of each stage in years divided by the doubling time of the GWP/person during that stage. It seems that, historically, the length of time which each stage has lasted has been roughly proportional to (and several times) the doubling time for the average income during that stage. Also, this doubling time has *decreased* by a significant factor in each stage. Note that the ratio for the agricultural age is difficult to quantify, partly due to lack of data. If we use the most secure growth rate in GWP/person noted in Table 1, that for the most recent period (which is actually too high to represent the whole period), and take the length of the total age as 8,000 years, the ratio is about 7. In contrast, the length of the industrial age (1820 to 2000) was 180 years, about 3.3 times its overall doubling time of 54 years (for GWP/person). At 4 percent, the doubling time for personal income in the digital age would be 18 years. If the digital age were to last 35 years, as suggested above, the ratio of the length of the age to its doubling time would be about two.

In summary: our main hypothesis is that each successive stage *overlays* the previous stage with a new technology that *multiplies* the summed technologies of earlier stages. It is this multiplicative effect that is hypothesized to be the dynamic underlying the large increases in growth rates with each successive age seen in Table 1 and predicted in Table 2. Figure 2 presents an overview of human development to 2050 based on Table 2 projections. This should be compared to Figure 1, which was based on a more conservative projection using an extrapolation of actual growth over the last 13 years.

Projection to 2050

If the projections in Table 2 and Figure 2 are valid, by 2050 the average person on this planet will have a standard of living that greatly exceeds that of people in the most developed countries today. These projections are, of course, for [Real GWP], i.e., a GWP adjusted to take into account non-monetary and quality improvements in life (discussed in the next chapter). However, even the GWP as traditionally calculated is expected to exceed the estimate used for Figure 1, due the acceleration of technology and underlying growth rates typical of the digital and following stages (outlined in Table 2).

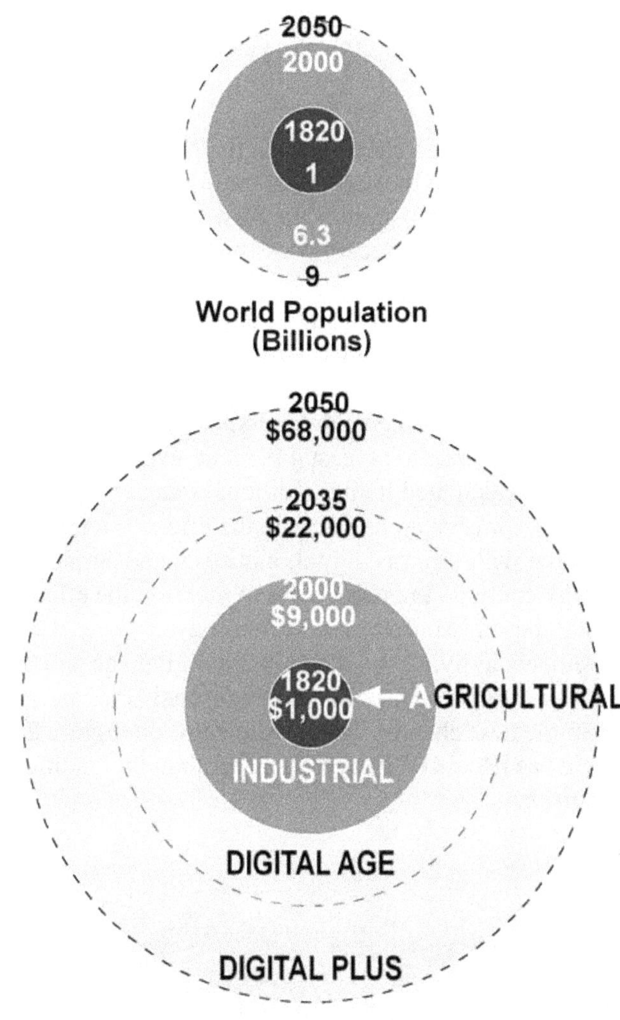

Average Annual Personal Income
(2012 U.S. Dollars per year)

Figure 2. Probable Human Development to 2050 CE
(For *Real GWP*, using Table 2 projections)

The projected effective average income of nearly $70,000 per year by 2050 is remarkable. This is an average income, of course, and I am mindful of the recent warnings that the gap between rich and poor is increasing, and the predictions that automation will only exacerbate this tendency. However, it is most likely that even the poorest third of people on a planet with an average income of $70,000 per year are going to be well off compared to the poorest third today where the average income is only $13,000 per year - less than *one fifth* as much!

There is considerable difference between the projections in Table 2 and Figure 2 compared to those of most traditional economists (e.g., Nordhaus, 2008, 2013). Most predict a growth rate of GWP of about 2 percent per year, based on recent performance. The estimates here are higher for the following reasons:

- The digital age has just begun and the growth rate of GWP/person is already moving above 3 percent (e.g., as predicted for 2014 by the World Monetary Fund).
- Comparison with the takeoff in the industrial age suggests that the rate will soon rise to at least 4 percent, even for the GWP/person calculated in the traditional manner.
- Traditional estimates do not include the effect of technological layering that will occur as digital, industrial and agricultural technology continue to interact, not to mention the effects of additional, possibly unforeseen, technology,
- Improvements in living standards in the digital age and beyond will exceed estimates based on the traditional GWP/person, as shown in the next chapter. That is, the values in Figure 2 are to be interpreted as [Real GWP]/person rather than the traditional GWP/person.

6. PROGRESS IN THE DIGITAL AGE MEANS MORE THAN GWP

The gross world product (GWP) does not provide an accurate estimate of the standard of living, particularly in the digital age and beyond. An increase above the traditional GWP/person of about $10,000 per year is estimated for improvements of life provided by the combination of digital photography and a mobile phone, and similar improvements. It is suggested that the GWP itself be formally replaced by [Real GWP].

Limitations of Using GWP as a Measure of Living Standard

As noted above, the gross world product (GWP) per person is the most often used single measure of the standard of living of the average person on the planet (e.g., see the excellent projections by Volk, 2010). As even the originators of the measure[31] (and others recently) have pointed out, however, it is not correct to use GWP as the sole measure of prosperity, nor to use GWP/person as an unqualified measure of personal income. For example, if there is an increasing disparity between rich and poor, the average personal income may be going up while the *median* (middle) income is going down.[32]

Even more significant problems include the fact that GWP does not include free services, does not capture quality changes and includes, for example, both the costs to degrade the environment and the costs to clean it up (e.g., Fleurbaey and Blanchet, 2013; Stiglitz et al., 2010). For these reasons, some economists (e.g., Nordhaus, 2013; Stigliltz et al, 2010)) prefer to use *consumption* rather than GWP/person as an index of living standard. However, this measure traditionally does not count free or quality improvements either. It is not the point of this chapter to make a quantitative estimate, but just to point out through relevant examples how using such measures may cause us to seriously underestimate some aspects of progress, especially in estimating people's well-being and enjoyment of life in the digital age and beyond.[33] That is, the possibility to be explored here is that the value of (some) digital technology in improving people's lives significantly exceeds its cost and may provide additional economic and environmental advantages compared to technological innovations characteristic of previous ages.

This is important for the following reason. Any serious prediction for the future must make some assumptions about the relationship between energy use and GWP, and how this relationship will change (hopefully improve) with time. The projected use of energy is used to obtain an estimate of carbon emissions, for example, and the GWP/person estimate is used as a proxy for the average standard of living.

Such a calculation connects the standard of living enjoyed by the average person on the planet to the global climate change caused by, and suffered by, the same person. This method of estimating falls short, however, if the use of digital technology gives significantly more value than its cost, that is, when the new activities involved cost significantly less in overall emissions than the activities that were replaced. If so, it would be possible to obtain more life satisfaction at a lower cost in emissions, with less deleterious climate change. If people don't know this, they will tend to overemphasize the fact that their personal income or country's GNP has grown slowly. They will lament that things are generally getting worse, and that their children will not have lives as good as they did. This idea is depressing to both child and parent, of course, even if demonstratively false (see Ridley, 2010).

Even a decrease in the GWP/person may not accurately reflect a decline in living conditions, rather, it may hide an improvement in living and in the cost of activities in terms of emissions. For example, consider how GWP would (will) be changed if (when) people begin to find it more enjoyable, instead of taking trips, to watch any of the many excellent nature and travel programs now available on digital television. Imagine how even more compelling it will be to stay home when programs are available in 3D with full color and immersive sound, and when the cost and inconvenience of air travel only increase. The cost of television is now on the order of hundreds of dollars per year, while the cost of the equivalent trips would be in the tens of thousands or more. Such a substitution would cause the GWP, as usually calculated, to *decrease* by the difference. The carbon emissions caused by television viewing also are clearly much less than those caused by a proportionate amount of air travel (which is a particularly problematic source of emissions).

A moving statistic: the number of younger people obtaining driving licenses and the amount this group travels by car was recently found to be decreasing for the first time ever in the U.S. Researchers speculate that people travel less by car for several reasons, all connected with the internet: 30 per cent of people now telecommute, people feel connected to others by interaction via internet and don't need to physically travel as much for social contact, people are doing more and more shopping by internet, and when they do travel, people tend to go by public transportation so that they can continue to use the internet![34] Again, this decrease in automobile sales and use means a lower GWP, but obviously the change is beneficial for the planet.

Most of the advances in digital technology are provided free to the average user, and free services are not counted in GWP. The internet in this sense is like the libraries of previous eras. However, the internet is much more than a free electronic library. It is "an electronic library bigger and better than any other library that has ever existed or even been contemplated by humans." It is also an electronic store, but "the biggest, best store ever, where you can buy anything from anywhere, based on reviews by other buyers, at a discount, and ... have it delivered tomorrow."[35] The GWP does not count the huge improvement in human welfare that is potentially made available in these ways. Nor does it include the provision of quality free services like Google, Facebook, Twitter, Wikipedia and YouTube except by their advertising revenue streams.

GWP not only doesn't count free services, GWP is *reduced* when products of improved quality but lower cost replace older products. Joseph Stiglitz et al. (2010) comment, regarding the GWP, "Capturing quality is a tremendous challenge..." Economist Tyler Cowen (2013) has noted that slower growth in health care costs is a boon to government and business, but will subtract from growth in GWP; also that revenues are now lower for record companies but more music is more available at lower prices, benefiting the consumer but lowering GWP.

A Typical Calculation of Equivalent Cost

One of the digital technologies that has clearly improved many aspects of life and business in present society is the combination of the digital camera and the mobile phone. This feature has many important uses in business, but here we consider just the enhancement of life for people who are now able, in conjunction with email and other internet services, to easily share photos and videos with friends and relatives. My experience on a road trip through the American Southwest may furnish a concrete example of the potential savings, or equivalent increase in life satisfaction. I found my enjoyment of the trip was considerably enhanced by taking pictures with my new smart phone and emailing them to friends and relatives. Each day, I would take 10-20 photos. Each night, I would review the photos in my phone, select and edit several of them, and share them by attaching them to an email sent simultaneously to a group list. This is not at all an usual activity today, by the way, judging from Facebook pages and emails I have seen.

One way to get an idea of the value of such an activity is to estimate the cost in time and money of the activity if one were to do it with what had been available previously. For specificity, let's assume a person took 10-20 digital photos each day using a mobile phone, viewed and edited them each evening, picked 3 of them to send via an email list to 5 friends and relatives, and did this for one week, for a total of 100 photos shared (20 to each person). Table 3 presents a summary of the amount saved using digital technology by comparing its cost to what would be the *additional* cost to do the equivalent activity with a 35 mm film camera, at present prices and wages. Costs to purchase each device were not included in this comparison. Similarly, since it takes roughly the same amount of time to take or to review pictures with either device, no difference in money or time was ascribed to these activities. The total additional costs to take and develop 4 rolls of film (96 exposures) and to make and mail copies of 20 pictures to five people is estimated to be $100 in direct costs plus 6 1/2 hours in time.[36] With leisure time rated at $15 per hour,[37] the total cost of doing things using a film camera and snail mail would then have been an additional $200. Note also that it would have taken several weeks longer, as well, lessening the immediacy of the experience for all participants.

Looked at another way, the advent of digital technology improved this person's life to the extent of about $200 per week, while at the same time *removing* about $100 per week from the calculation of the GWP. That is, if the person had done the same activities the old way, he would have had to spend $100 out of pocket for buying film, developing, copying, and mailing the photos. All of these expenditures would have been counted in the GWP for that year. In contrast, taking and sending many copies of digital photos is essentially free; there is *no* contribution to GWP from this activity. The only item which could potentially be counted toward GWP would be the original purchase of the mobile phone, which would at least partly be offset by the cost of the film camera, any remainder appropriately being included in telephone expense.

Table 3. Amounts Saved Using Digital Camera and Email

Activities Replaced:	Amounts Saved: Money	Amounts Saved: Time
Drive to store, buy 4 rolls of 35 mm film, total 96 exposures	$19	1/2 hour*
Take pictures	—	—
Drop off film to be developed	$40	1/2 hour*
Pick up developed photos		1/2 hour*
Choose **20** pictures to send.	—	—
Make 5 copies of each photo selected, total of **100** copies	$28	1 hour*
Several weeks after the trip, write notes to 5 different people and address envelopes	$5	3 hours
Mail 20 prints to each person	$8	1 hour*
Totals	$100	6 1/2 hours
	$200	

* Includes travel and shopping time:
Additional costs for transportation not included.

In addition, the increase in leisure time of 6 1/2 hours using a digital camera would not be included in the GWP for that year. This means that an amount of about $200 per week or $10,000 per year would be *missing* from the accounting of GWP for this person. If the same is true for many other people and for other activities (and it seems to be), then in this way the GWP/person would significantly underestimate the true level of life satisfaction in the digital age.

You may object that no one in earlier days would have sent so many photos to friends. Not true, we actually did much the same thing in the old days, but accomplished it in a different manner. For example, my father was a typical amateur photographer. On a vacation, he would take many photos with 35 mm slide film. After, he would gather together friends and relatives for a big slide show, where we were usually treated to *many* more than 20 photos. Later, I would find myself doing much the same thing but with an 8 mm movie camera. I think the method today, where you can look at photos sent online to a group, or posted on Facebook, or not look at them as desired, is a lot less time consuming and less expensive for everyone.

Therefore, the cost comparison in Table 3 was properly done, with the "old way" estimate being made so as to duplicate, as much as possible, the typical way it is done today. To the extent that the typical digital user *may* share many more photos with more people (e.g., by email, on Facebook or on another personal web site) than would have been done previously, sharing digital photos is a *new* activity, not one which simply replicates or replaces an earlier process. The internet has created many similar activities which are completely new activities, activities for which there was no previous equivalent in the industrial age (Reese, 2013). The novelty of the many new activities available makes it even more difficult to estimate quantitatively the improvement in living that the internet and other digital technology has enabled, and no further attempt will be made in that direction here.

You may obtain a fairly good estimate for yourself by asking, "How much money would I have to be handed each year to make it worthwhile to give up, for that entire year, *all* of my activities that are enabled by digital technology?" Think carefully about all aspects of your life that may involve digital technology. Please take a moment to answer this question before you read further. [38]

.....

What was your answer? Whatever your answer was, if a person would require that much money to give up an important aspect of his current life, he is, in some sense, that much wealthier than he would be otherwise. *None* of this equivalent income is presently included in personal income estimates.

7. GLOBAL WARMING AND OUR GLOBAL SOCIETY

To cope with global warming in the digital age, given the accelerating rate of technological development, it seems most useful to make the *assumption* that, by 2035, non-polluting energy sources will be widely available and inexpensive compared to fossil fuels. More draconian measures to curb carbon pollution should be withheld until the situation is reevaluated then. In the meantime, efforts should be concentrated on improvements in power distribution and energy storage, in energy use efficiency, in coping with rising seas, and in predicting consequences of climate change.

Global Warming and Automation

A thorough review of the current literature (see Bibliography) leads to the conclusion that the two most important and difficult issues facing the human race today involve 1) global warming, particularly human-caused climate change, and 2) the effects of increasing machine abilities, particularly the near certainty that most current human jobs will soon be performed by machines. These two issues seem particularly difficult because for both:

- The processes are already well underway, and have considerable inertia.
- There are very long-term and largely irreversible consequences.
- To avoid the problem completely, progress of our civilization would have to be stopped completely (or even regress).
- In spite of the risks, there are many potential benefits to continuing our technological/social progress.
- These are complex problems spanning many areas of the technological and social sciences.
- These are new problems facing the human race (at least to some extent) and no one can confidently predict outcomes.
- Both problems are likely to be affected significantly by events over the next 20-30 years.
- Unfortunately, the coming technological changes and their implications are not predictable at even this short time scale.
- The two problems are essentially intertwined but mostly considered separately, primarily because they involve wildly disparate fields of knowledge.

The advent of full automation and the consequences for human society are considered further in Chapter 9, while the specific dangers posed by machine superintelligence are explored in Chapter 10. The problem of global warming will be discussed below in the following order: what we know for sure to date; what is most likely to happen in the next 35 years; what should be done about the problem over the next 20 years; and dangers of the path recommended here.

What We Know For Sure to Date

The worldwide scientific findings have been summarized in the most recent report[39] from the International Panel on Climate Change (IPCC). Because this is a consensus report from a large group of scientists, all its findings are presented as having varying degrees of certainty. Unfortunately, this has been seized upon by some who either ignorantly or willfully are misrepresenting the findings. When you can get a large group of scientists, as cautious, skeptical and independent as scientists tend to be, to all agree along with their governments that something is "extremely likely," I contend it is something that we can *assume* ***is*** certain for all practical purposes. In that vein, the important findings from the IPCC and other reports are summarized below.

Careful measurements at Mauna Loa, among many other places, show that the carbon dioxide (CO_2) levels in our atmosphere have rapidly increased in recent years. Before the industrial Revolution, the CO_2 level had held steady at 280 parts per million (ppm) for thousands of years. Since 1840, however, it has increased to reach nearly 400 ppm, and its rate of growth is increasing. As has been adequately discussed elsewhere[40], CO_2 is a greenhouse gas, trapping heat in our atmosphere that would otherwise be radiated out to space. Methane (CH_4) is even a more effective greenhouse gas than CO_2, and its concentration has also increased substantially, currently accounting for 60% again as much warming as CO_2 by itself. To date, the net result of the increase in greenhouse gasses has been a "small" but significant increase in the average temperature of the planet, about 0.8 °C (1.4 °F).[41] This is an average increase, a result of measurements taken over the whole planet.

Unfortunately, the effect on our planet of the increased heat retention is not just a rise in average global temperature. The air temperature at the poles has increased about twice the warming averaged over the whole planet. This has caused an increased melting of permafrost in

Arctic regions and loss of glacier ice. In addition, over 60% of the total energy increase absorbed by the planet has gone to warming the top 700 meters of the ocean. The expansion of the ocean due to this heating plus the increased melting of ice on land is has caused the average sea level to increase by 19 cm (7.4 inches) to date.[42]

Other effects of global warming have varied considerably by region. Most regions have had, as one might expect, a significant increase in the number of hot days and nights, while some regions have had increased rainfall and others have not. In a few areas an increase of drought is suspected of being a result of global warming, but the evidence is not yet conclusive. Overall, it is concluded that global warming to date has led to a significant *increase in climate and weather variability.*[43]

Human activities are the only believable source for the increased levels of carbon dioxide and methane in the atmosphere. The fact that this result continues to be disputed by a few is partly due to the complexity of the carbon cycles on our planet, but primarily due to activities of a small group of scientists and others *paid* to cast doubt, paid by organizations formed by factions who (erroneously) thought that it was in their interest to confuse the issue. Obviously, some oil and gas companies might think that they had reason to fear this finding, and accusations have been made in their direction. It is an incredible story, but reportedly some *tobacco* companies also secretly formed organizations designed to cast doubt on climate science as a way to try to change public perception about the validity of scientific findings in general. Their goal was to try to avoid public action regarding the effects of second-hand smoke (Monbiot, 2007). It is a sad irony that the issue of second-hand smoke has been mostly resolved by banning of smoking in restaurants, the work place, etc. (at least in most of the U.S.) whereas the fact of human-caused climate change is still disputed by some people but not, however, by the *overwhelming* majority of experts in the field (Hanson, 2009). In the final analysis, the damage done to the public understanding of (and trust in) science may be the worst outcome of this whole sad debacle. In any event, the 2013 IPCC report now carries the statement that it is *extremely likely* that global warming is due to human activities, which I translate as *certain.*

What We Can Expect Over the Next 35 Years

The excess carbon dioxide already in the atmosphere will persist for a relatively long time, and will continue to cause additional global warming even if all human emission-causing activities were suddenly stopped (Nordhaus, 2013; Stager, 2012). More to the point, the current trajectory of human emission production is not likely to change drastically in the near future. The amount of warming over the next twenty years is therefore essentially determined. Future warming, beyond more than about 25 years from now, depends in a complex way on economic and social changes, technological development, and decisions made by governments, not to mention complex responses of the whole climate and ecosystem. The amount of warming expected in the future is generally estimated by comparison of the outputs of a number of different global climate models (GCMs) by different groups, which are the reports that IPCC depends on for their predictions.

Unfortunately, even assuming a particular pattern of greenhouse gas emission in the future, existing GCMs give a very wide range of responses. Consider, for example, the predicted responses in climate resulting for the "worst case:" that with no intervention, or "business as usual" in terms of economic growth and emissions. For this case, the predicted temperature rise compared to today, *as soon as 2050*, ranges from 1 °C to 2 °C (within 90% probability limits). Including the potential range of possible emission patterns, the overall range in outcome at 2050 ranges from essentially no change to a 2 °C warming, a huge uncertainty!

The main problem for scientific prediction is the unknown effect of clouds on weather, including the interaction of particulate pollution and clouds. David Orrell (2007) raised the alarm over this problem with the publication of the results of his doctoral thesis. In this work, he mathematically analyzed a large number of outcomes from typical GCM programs with minor changes in initial conditions or assumptions. He showed that the resulting outputs did *not* have the mathematical form that they would have if they were the result of chaos in the climate system or uncertainty in initial conditions, as generally had been (and is often still) assumed. Rather, the variation in response was apparently due to inadequate, non-physically based parameterization of the effects of clouds. Because cloud parameterization is not based on fundamental physical knowledge, for each model run the parameters are adjusted until the model output reproduces the climate up to the present date. This means that the real weather could well depart significantly from the predicted weather as soon as the physical parameters (e.g., CO_2 concentration) depart significantly from historical values. Orrell suggests that it is not even adequate to rely on the *median* output of a number of runs in order to arrive at a reasonable prediction, as the IPCC generally does (and as we do below), because the lack of knowledge of the physics of cloud formation means that the *actual* climate response could even lie outside the range of model outputs, especially as distance into the future increases.

The most recent IPCC report acknowledges that the uncertainty in cloud dynamics, including the interaction with particles, contributes nearly as much uncertainty in the predicted response as does the total effect of carbon dioxide (Figure SPM.5). Nonetheless, the IPCC results *can* give some useful ideas about the most likely responses we can expect, at least for the immediate future before conditions diverge too much from historical records. We therefore summarize some of the median responses and predictions from the last IPCC report.

The median prediction of the most optimistic IPCC model, that with the lowest total emissions (RCP2.6), predicts global air temperatures increasing another 0.7 °C by 2050, and then being stable from that point onward. A more likely model may be the second-worst case illustrated (RCP6.0), which predicts a median increase over present temperatures of slightly less than 1°C by 2050. This model also predicts an additional rise in sea level of 17 cm (6.6 in.), as the median increase from the present level. Even the worst case illustrated by the IPCC (RCP8.0) does not result in significantly more serious outcomes at 2050. However, worst-case outcomes *do* become *much* more serious by 2100, if no intervention is made.

The IPCC report also predicts that extreme weather events will be expected to continue to increase in frequency, at least through 2050. This will include increased rainfall in many areas, increased drought in some areas, record levels of hot days and nights in most areas and regional heat waves in specific areas (e.g., southern Europe). It is likely that there will be an increase in the incidence and/or magnitude of extreme high sea levels, with severe storms causing increased flooding of low coastal areas. It is also predicted that the total monsoonal rainfall will increase, but may become more erratic.

The results of global warming will not be all negative. A decrease of summer Arctic sea ice by 2050 is expected to open new ocean transport routes and areas for mineral extraction. Further, predicted changes in rainfall and growing season suggest that climate change may open new areas for agriculture, such as the Canadian and Russian north, even as other areas become less suitable.

Overall, however, the effects of global warming are predicted to be largely negative, including overall reductions in food production, increased damage from storms, a huge loss of species diversity and a significant rise in sea level, by itself expected to cause a substantial loss of human habitat, agricultural area and wetlands. The net potential financial cost of these damages is discussed next.

What Should We Do About It in the Immediate Future?

In summary, I think that the scientific data and analysis are clear enough, especially with the last IPCC report, that the following statement will be agreed to by all but the most contrarian deniers: Human-caused greenhouse gas emissions have caused significant changes in global climate to date and will cause increasing changes in the future.

What we should do about it, however, depends very much on how serious we expect the results to be. The amount of damage that has been and will be caused by global warming is *much* harder to estimate, for a number of reasons.

- There are potentially large non-monetary damages, and comparing the cost of such damages depends on valuation processes which vary wildly according to individual values. What is the existence of a particular species worth, or a particular society? How much have we lost if a beautiful city drowns?[44]
- The future amounts of greenhouse gas emitted into the atmosphere depend in a complex way on choices made by individuals and societies around the world, which depend, of course, on the choices available.
- As noted above, the consequences for the climate of even a given amount of greenhouse gas emission cannot now be accurately predicted, mostly because the effect of clouds and of the complex interactions of particulate emission with clouds are not adequately known. The possible presence of "tipping point" transitions (discussed below) further complicates damage estimates.

- As noted in earlier chapters, the technological innovations to come and implications of these changes are unknown and unknowable for times much more than a decade into the future. New technologies might reduce energy need, provide cheap carbon-free energy sources or provide improved methods of reducing negative aspects of climate change.

As a consequence, the range of suggested solutions is huge. Those identified as "environmentalists" have typically valued ecosystems very highly, and from the beginning some have suggested draconian reductions in fossil fuel use, with potentially large (and largely unestimated) economic consequences.[45] At the other extreme are those who value economic progress (or coal production) so highly that they resist any restriction of fossil fuel use, to the extent that they have even attempted to deny the existence of the problem (e.g., Ridley, 2010).

In this debate, it may be noted that, while most "futurists" do not consider climate change in their projections, neither do most “environmentalists" take into account the possible development of technologies predicted by "futurists." This in spite of the fact that climate scientists *have* to be "futurists" because, in order to predict the effect of global warming, they have to have a mathematical model of future society to estimate the amount of CO_2 that will be released into the atmosphere. In practice, therefore, in order to make the quantitative estimates that are absolutely required for this kind of prediction, climate scientists and economists must make predictions about the rate of carbon emission in future societies (worldwide) based on an extrapolation of current trends (Nordhaus, 2013). They cannot, obviously, include the effect of *unforeseen* technology in such calculations! While they cannot incorporate such changes in their quantitative predictions, however, environmentalists *could* at least include the *possibility* of such changes in their suggested solutions which has generally not been done (but *is* attempted below).

Somewhere in the middle of this debate are the economists, who attempt to balance society's desire for improved living conditions against future damage caused by global warming. While these estimates have largely ignored some of the very difficult to quantify costs of species, habitat, and human lifestyle loss, it is worth reporting the monetary estimates for the damages that *are* included to get some rough idea of the overall magnitude. For example, Nordhaus (2008) has estimated the total economic damages caused by climate change with no intervention to be $25.7 trillion (in 2012 dollars) over the next 250 years. By comparison, the gross world product (GWP) is currently $72 trillion per year. By the year 2100, the damages from unrestrained environmental damage are estimated to amount to 2.5% of the projected GWP/year.

The Nordhaus model estimated that the total cost (damages plus mitigation) of global warming could be reduced to $20 trillion by the most efficient intervention. That is, the human race could save itself a total of $5.7 trillion by encouraging a change to alternative fuels over time, decreasing the carbon dioxide emission below that with no intervention. One straightforward way to accomplish this would be imposition of a tax on the amount of carbon dioxide produced by a fuel. The intent of this "carbon tax" is to cause carbon emitters to pay part of the actual costs caused by these emissions, and to encourage development and use of alternative, non-carbon producing energy sources. The tax recommended by Nordhaus (2008, 2013) would result in an increase of about 8% in the price of gasoline in 2015, for example.

It is important to understand that, in these calculations, future costs are discounted at a rate reflecting the return on capital. Nordhaus assumes for his model that the rate of return on capitol is a historically justified minimum value of 4% per year. This means that an expense of $104 a year from now can be countered by an investment of $100 today. Differences in the assumed discount rate have a *large* effect on the cost of future damages in today's dollars, and explain most of the huge variation in published estimates.[46] A lower or zero discount rate, assumed by some environmentalists, leads to much higher estimated costs from global warming.

There are several important caveats and conclusions relative to the Nordhaus model study.

- Technological innovation is an *input* into the model. The model does not include technological development which is encouraged or driven by economic forces - the most valuable technology! When similar transitions have been legislated (e.g., improved gas mileage) need-driven technology has historically lowered net costs to *half* of the amounts estimated in advance (Hansen, 2009).
- Given his optimum minimum estimate of the cost of climate change, Nordhaus concludes that the economic benefit of a low-cost and environmentally benign backstop technology (as an

alternative to fossil fuels) has a present value of *$20 trillion dollars*. Any companies or inventors out there interested?[47]

- Uncertainty of the costs and consequences of global warming suggests that *"a more rapid resolution of (this) uncertainty implies that it may be beneficial to impose less costly restraints until the exact nature of future consequences is revealed"* (Nordhaus, 2013, p. 193).

These considerations lead me to suggest a somewhat radical suggestion, at least from a person who considers himself an environmentalist. I have been forced to the proposal to follow by the following line of reasoning.

- As shown in earlier chapters, the rate of technological change is already high and is likely to increase even more rapidly in the coming decades. Technological innovations almost certainly will include much smarter and faster computers, which will, among many other things, improve measurement and prediction in climate models.
- The price of alternative power has been dropping exponentially relative to carbon fuels for many decades. Already, wind power is competitive with the best natural gas plants, and much cheaper than coal. Utility-scale solar and geothermal power will soon both be much cheaper than any carbon fuel.[48] When this happens, it will not be necessary to encourage developing countries or others to invest in alternative power sources, they will simply do so because it will be so much cheaper. Note that no new, unexpected technological breakthroughs are assumed here, just a continuation of the historical trends in development As discussed in Chapter 11, the probable advent of new, unexpected technology only increases the likelihood that alternative and inexpensive power sources and energy storage methods will be developed by 2035.
- As discussed above, computer intelligence in particular is predicted to become significantly greater than human intelligence by 2035, a little more than 20 years away, creating yet another jump in the technological change rate. Even if this particular technology does not materialize fully, there are several other probable technological areas that should produce massive technological changes by that time.

Given all these above considerations, I suggest that will be advantageous for us to *assume* that a sufficient variety of

alternative energy sources will become significantly cheaper than fossil fuels within twenty years from now, with only minimal intervention. Governmental and other interventions could then be focused on those improvements needed to maximize benefits from this result, and in particular, would include measures to upgrade power grid and power storage capacities. Also appropriate will be measures encouraging increased energy efficiency, provided they do not significantly interfere with long-term economic progress. Funding research on understanding global weather processes as well as ecological and social consequences of global warming will continue to be very important, of course. (Desirable interventions are discussed in more detail below.)

Delaying implementation of other, more draconian measures until 2035 will not cause significant differences in the outcomes for global warming expected by that date. It may even be that alternative energy sources will be enough cheaper than fossil fuels by then that *more draconian measures may not be needed.* The global warming problem may be largely solved, or well on the way to being solved, by then.

Dangers of the Recommended Path

My friends and colleagues who are climate scientists and environmentalists should and do raise some serious objections at this point. Potential problems with the proposal above are examined in turn below.

Alternative energy sources might not become sufficiently attractive twenty years from now. The chances of this appear near zero. Wind and geothermal power already cost less per kilowatt-hour than the cheapest form of fossil fuel (gas combined cycle) and much less than coal. Utility scale solar photovoltaic energy costs are already comparable to coal. The National Renewable Energy Laboratory conservatively estimates that prices of *all* alternative energy sources will continue to become more attractive, reaching half their current values (in constant dollars) over the next 15-20 years. This estimate includes advanced technologies currently being developed for geothermal energy which are expected to make it *much* more widely available, providing a badly needed 24-hour background energy source.[49] This means that alternative fuel sources will become more attractive compared to fossil fuels even if these don't rise in price in the future. As noted above, these attractive estimates do not include additional advantages of unknown technological improvements for any alternative sources, that is, the development of improvements which are very likely to occur given the history of these fields and the present technological forecast. I do not think we want to totally discount the idea of the emergence of a "backstop" technology that will provide a cheap and environmentally friendly alternative to fossil fuels (neither do I think we should count on it). The real stumbling block in this context would be a substantial drop in the price of coal, which *would* mandate an immediate imposition of a carbon tax.

The negative effects of global warming might be much more serious than currently estimated. There are two major issues which could lead to this outcome: inadequacies in general global climate models, and the possible triggering of "tipping points" in regional climate systems.

A) The actual effects of global warming might be much worse than our current models predict. As mentioned above, the effects of clouds and their interactions with particulate matter are very poorly understood. For this reason alone, IPCC projections become increasingly uncertain as time goes on. For 2035, however, the total uncertainty is estimated to be only 1 °C, with the most likely rise in average temperature being about 1 °C over the present value.

It seems unlikely, therefore, that there will be any completely unexpected dynamics in our climate system before 2035. Certainly, no logic or paleontological studies suggest a sudden catastrophe could result from such a small change. By 2035, we should be able to improve our forecasting considerably - as a result of 20 years more data, much faster computers, and (hopefully) a much improved understanding of cloud/particle dynamics. Then we will be better able to assess and deal with potential future problems. It should be stressed that a reassessment by that date will be an important and necessary action.

A related issue here is that the potential damage to agriculture and other climate-sensitive production could be much worse than predicted. However, the effects will be severe only in undeveloped areas. While unfortunate, of course, this damage is unlikely to significantly disrupt technological advance in the next twenty years, because that primarily will occur in developed economies and agriculture is a very small part of these economies (Nordhaus, 2013).

B) Global warming may trigger "tipping" transitions that cause catastrophic damage. A number of climate scientists have raised alarms about the possibility that system feedbacks in some parts of the weather system may be such that the subsystem will undergo a sudden and potentially disastrous change as the global temperature rises.[50] Going through a tipping point implies a certain amount of irreversibility (as in tipping over a glass of water). Both the old state and new state are stable states, which require large perturbations to move away from, but once the transition is made to a new state, the system tends to stay there. The physics of suggested tipping point transitions are poorly understood, however, and it has not been generally possible to include them in global climate models to date.

The only tipping point considered in detail in the IPCC report is the possible collapse of the Atlantic Meridional Overturning Circulation, aka the "Gulf Stream" ocean current which keeps Europe and Northeastern America warmer than it would be otherwise. A shutdown or decrease of this current is thought by some to have been the cause of the "little ice age" that affected Europe and the East Coast of North America from 1300-1850 CE (Fagen, 2000). However, current measurements do not reveal a significant weakening of the current, and the IPCC report concludes that the current is unlikely to collapse before 2100. Note also, that a decrease in temperature might actually be welcomed by that time.

The other major possible weather catastrophe mentioned in the IPCC report is the complete failure of the monsoon, and this also is not expected to occur before 2050. There are no other known tipping points predicted to occur until the total global temperature rise exceeds 2 °C. Since the total rise by 2035 is almost certain to be less than this (if only slightly), it seems likely that we do not have to worry about a tipping point transition in the next two decades.[51]

The disappearance of Artic sea ice during the summer is often suggested as the first tipping point we will see. However, this event, while it will be a significant milestone, and one which will probably occur by 2060, is not a genuine "tipping point." The climate, even locally, will be about the same the first summer that there is no ice as it was the summer before. It has been suggested that an Arctic tipping point *will* be reached via a positive (runaway) feedback loop when the increased temperature melts the Siberian and Canadian permafrost, the permafrost melt releases methane, and the additional methane causes a further increase in global temperature, and so on. While certainly there is some permafrost melting due to global temperature increase, there is not yet rapid enough melting to cause a *runaway* feedback temperature rise.

Governments may pass ill-advised legislation which disrupts economies but doesn't effectively counter global warming. Two examples of this come to mind. The first is artificially restricting carbon emissions through ineffective international agreements, in the way that the Kyoto Protocol has attempted do. A recent experience of mine brought home a core difficulty with this agreement. I was in the Australian bush on a research project. We were driving through a beautiful Eucalyptus forest when the road suddenly dead-ended in a thick evergreen forest, dark and impenetrable, and completely out of place in Australia. Upon inquiring, I was told that a Japanese company had cut down the Eucalyptus forest and planted another forest under a provision of the Kyoto agreement.

Think of this: An international agreement intended to reduce emissions had led to a company cutting down a very beautiful, perfectly functional forest, one that supported a range of wildlife, in order to plant an unnatural, ugly forest that would provide no home for wildlife and which did not actually sequester any significant additional carbon, all so the company could get a "carbon offset" and continue to pollute the atmosphere exactly as it had before. Any agreement calling for artificial restrictions will have such loopholes, of course, which is what leads to such unproductive behavior.

Equally to the point, international agreements like Kyoto are very difficult to modify to adjust to the fast changes we expect in the interacting social/technological and global warming systems. It is not within our scope to do a detailed analysis of the Kyoto Protocol, except to point out that reviews by economists and climate scientists are generally negative (e.g., Hansen, 2009; Nordhaus, 2013). The bottom line is that, even if this protocol were to be followed by all nations, it would not result in a significant reduction of CO_2 emissions. It has not done so to date.[52]

Biofuel production provides a second example of ill-advised government intervention. As Matt Ridley (2010) argues,[53]

> Not even Jonathan Swift would dare to write a satire in which politicians argued that in a world where species are vanishing and more than a billion people are barely able to afford to eat it would somehow be good for the planet to clear rainforests to grow palm oil, or give up food-crop land to grow biofuels, solely so that people could burn fuel derived from carbohydrate rather than hydrocarbons in their cars, thus driving up the price of food for the poor.

Even if derived from non-food sources, biofuel production causes sufficient carbon and environmental costs to make it the last choice for a major source of alternative energy (Worldwatch Institute, 2013). There are plenty of better alternatives already, as summarized above.

Fortunately, neither of the two examples above, even if they do constitute serious errors, have the potential to seriously disrupt economic and technological progress over the next several decades.

Our main conclusion is then: Given the rapid development of the world economy suggested by our historical analysis and present trajectory, and given the current and projected developments in alternative energy sources, current plans for coping with global warming should be made with the *assumption* that, by 2035, alternative energy sources will be available which are significantly less expensive than fossil fuel energy. Alternative sources will then naturally tend to displace fossil fuel use with minimal further intervention, except as will be required for power grid and related changes.

At this point, in other words, I essentially agree with Matt Ridley (2010) when he warns against taking expensive measures now which may limit the improvement in future prosperity of (especially poorer) people in the world. He sums his arguments thus: "*In short, a warmer and richer world will be more likely to improve the well-being of both human beings and ecosystems than a cooler but poorer one.*" Our suggestion is to take on this point of view, to some extent, but to reevaluate it in twenty years to see if more severe measures might be required at that point.

In the meantime, it is suggested that we avoid draconian measures which might significantly limit social, economic and especially technological progress over the next few decades. *This does not mean doing nothing.* There are many steps that we can and should take to prepare the ground for the coming transition to alternative energy, steps that will not significantly interfere with (and may assist) social and technological development.

Recommended Steps Over the Next 20 Years

Steps that we do choose to take must take *enormous* uncertainties into account. Inadequacies in precise knowledge of dynamics of the weather system, especially clouds and tipping point transitions, along with current limitations of computer modeling mean that future results of global warming are very poorly predictable at present. The amount of temperature change predicted only 35 years from now lies between no change and 2 °C, which is a very large change in terms of the effect on the planet. Actual results could even be better or even worse than the model extremes.

As we have noted in previous chapters, social, economic and technological changes in society are uncertain more than a decade into the future, even *without* the complications of global warming. Therefore, the future of global warming must now be considered to be *equally uncertain* as that of our technological society. Our decisions about what to do now should be made with this in mind. Our choices now must be made on the understanding that we do not know the future of *either* our technological society or of global warming. Both systems are doubly uncertain because of the strong coupling between the two. Accordingly, it is suggested that all interventions be selected with a strong bias toward the "do no harm" dictum of medicine.[54]

1) The highest priority will be development and installation of a global power grid, including sufficient energy storage capacity, to make the best use of alternative energy sources. While some new technology would be useful to improve storage methods, most of this task can be accomplished by upgrading aging power systems to improve both capacity and control using technology already available. As most of these improvements will be useful no matter what evolves in either society or climate, this work can and should be started immediately and be planned to continue over the next two decades.

2) An increase in overall energy efficiency is another top goal. Overall energy efficiency compares the standard of living achieved in a society with the amount of carbon dioxide put into the atmosphere. The usual comparison is the dollar amount of GWP/person produced per unit of carbon release. However, note from the previous chapter that GWP/person does not appear to take into account many of the improvements of life currently experienced in the digital age. While a quantitative estimate is not presently available, therefore, at the least we could denote the improved estimate by the term [Real GWP/Person]. This goal should therefore be revised to maximizing the ratio of the [Real GWP/Person] to carbon emissions.

Increasing energy efficiency includes increasing the fraction of energy output from non-polluting power sources, decreasing energy transmission costs, decreasing the overall waste of energy through conservation at the point of use, including life style changes that reduce the need for energy, and increasing the efficiency of energy conversion to desired output (e.g., compact fluorescent light bulbs). Of course, the proper calculation of carbon energy efficiency must include all the fossil fuel contributions to the manufacture of all products involved.

Increased energy efficiency is an important goal because it is a "win-win" solution (e.g., Pearce, 2007; Volk, 2010). When increased energy efficiency gives an immediate benefit, it will be adopted naturally and no particular strategy is needed to encourage it. However, most energy efficiency improvements require some initial investment, i.e., most result in long-term but not short-term gain. In this case, governments will be called upon to implement legislation that encourages companies and individuals to take a longer view of their profit and loss statements. Immediate tax credits for increasing energy efficiency are a traditional and largely effective way, but not the only way, to provide such incentives. Appropriate regulations (e.g., auto mileage standards) can also encourage movements toward greater efficiency. Government funding should also be directed toward basic research that might lead to increased energy efficiency.

Some efficiency improvements may have a limited future. In encouraging changes, it is important to note that the savings from some energy efficiency improvements cannot be projected far into the future when the change rate becomes as high as it is predicted to become. That is, it will only make sense to ask people and businesses to make the investments in energy efficiency that pay off in a *decade* or less, given the rapidity of changes in technology and society. After all, a company is likely to be doing something very different ten years into the future, or it is likely to go out of existence by then. In that time, a new, more efficient product or power source may be developed that makes the previous one obsolete before its investment has paid off.

It would be very useful (some might say it was incumbent) on the U.S. to take the lead in helping the world meet the goal of increased energy efficiency. We produce twice the emissions that Europe and Japan do for a similar standard of living (e.g. Randers, 2012). We could afford to spend more than the few percent that we currently spend in this area, compared to the 51 percent of government spending now going to the military. If everyone in the world is better off, we are not likely to need as much military defense. Note further that energy efficiency improvements invented in the developed countries can allow undeveloped areas to leapfrog over intermediate steps in their development, resulting in less carbon emission as they develop.

3) Reduce methane leaks. Because the future is so uncertain, immediate efforts to reduce atmospheric methane would be very useful and should be given the highest priority now (Hansen, 2009). Methane (natural gas) is a major greenhouse gas, which the IPCC estimates is currently causing about 60% as much additional global warming as CO_2 does by itself. While it doesn't remain in the atmosphere nearly as long as carbon dioxide, its greenhouse effect is 30-70 times worse than CO_2, depending on the measure used to compare the two. Major sources of atmospheric methane are leaks from gas pipelines and underground sources (released by fracturing, mining or oil drilling), swamps, peat bogs, melting permafrost, landfills, agriculture and cattle. Capturing methane and reducing its loss is a win-win because the captured natural gas can be used as an energy source, one which produces the best carbon dioxide to energy ratio of any fossil fuel.[55] Unfortunately, while methane emission is very much a major problem, it hasn't gotten nearly as much attention as carbon dioxide; this is doubly unfortunate, as it would be much easier to remedy and cost virtually nothing economically.

4) Over the long term, practical ways to encourage a reduction in carbon dioxide production will probably need to be implemented. There have been a number of different proposals to encourage such reductions in fossil fuel use. The "carbon tax" suggested above seems straightforward enough to work and has the support of many environmentalists and economists (but not coal companies!). Nordhaus (2013) compares a carbon tax and a "cap and trade" system, noting that either can do the job if correctly and fairly implemented. To get a level playing field, and to allow the most natural and efficient economic development, the first step is to remove all subsides on fossil *and* alternative fuels. For the carbon tax, known pollutant fuels would then be taxed in approximate proportion to the amount of pollution attributable to them. The most appropriate way to accomplish this is a tax on the amount of carbon, based simply on the total number of carbon atoms in the product. The tax would necessarily rise over the years, and be large enough to discourage fossil fuel use and promote alternative fuels.

The money collected from the carbon tax could be divided by the number of adult citizens in each country and disbursed directly to each (the fee-and-dividend model supported by Hansen, 2009). Our suggestion would be to graduate the dividend according the person's income, or use it to provide support for unemployed persons. As we will see later, this idea will become increasingly useful as a way to keep consumer spending up *if* automation increases unemployment. Distributing the money according to the total population (i.e., including children) is not desirable because it would encourage people to have more children. From the discussion regarding the effects of automation in Chapter 9, it can be clearly seen that the alternative suggested method of distribution, a tax credit, would be a *very* poor idea because it would unfairly punish the increasing numbers of people who manage to find useful ways to live without traditional jobs, as well not providing them with income support.

The carbon tax, as suggested, would increase only gradually, so as not to disrupt economic growth. Even though such taxes would certainly be passed on to consumers in the form of higher fuel prices, imposing such a tax would not seriously disrupt economic growth as long as the money collected was returned to circulation in a useful way: as a dividend to each citizen or unemployed person and/or spent for mitigation or adaptation to carbon pollution including research and development of non-polluting power sources, and/or used in support of areas which also encourage economic growth (Acemoglu and Robinson, 2012) including expenditures on education, health, basic research, entrepreneurship support and other programs to benefit small businesses.[56]

There is not a consensus in favor of a carbon tax at this time, at least not in the U.S. Given that, I suggest the following process be considered: 1) Begin to implement other recommendations in this chapter; 2) Work on an international agreement on a *mechanism* for a carbon tax, with a start date about 2035; 3) start to impose a gradually increasing carbon tax at that time only if it was definitely determined to be needed to encourage the further shift away from greenhouse gas emission. Such a decision would be made by a group of international stakeholders as set up in the founding mechanism and would depend, of course, on the relative costs (unregulated) of fossil and alternative energy sources at the point of use.

5) Coping with sea level rise will be required. The rise in mean sea level so far has been 19 cm so far (7.4 inches), and by 2050 it is predicted to rise between 19 and 27 cm more (7.4 to 10.6 inches more). The net amount of rise in any area depends also on the vertical movement of land. Obviously, existing developments near sea level will need to be protected or abandoned, and new construction and land use rules should take current and predicted sea level changes into account.

6) Changes in life style allowed by the technology of the digital age that act to reduce carbon emissions should be encouraged by tax structures, or at least not discouraged, or at the very least, be acknowledged. On the national level, a correct accounting of the increase in [Real GWP] would be useful, so that people realize that they indeed are better off than their parents generation even if by traditional measures there has been little improvement and so that people can accurately compare their country's economy and lifestyle with others.

7) To maximize the possibility of technological and social innovation, governments around the world should be encouraged to reduce the outdated regulations and the corruption which limit competition and the development of new businesses. This is easier said than done, of course, and in many cases it will be technological developments or outside agencies that make an "end run" around the stultifying bureaucracies endemic in developing economies. For example, this occurred in Nigeria when the explosion of mobile phones simply made the "epileptic" government-controlled landline company obsolete, allowing people to start a wide range of informal businesses. However, it is noted that there are still significant barriers keeping Nigerians from taking their small businesses to the next stage, e.g., moving internationally, and this could still be improved either through internal change or external support from international organizations.[57]

8) Developed countries should stop constructing new coal power plants entirely, and, most important, encourage undeveloped countries to leapfrog this technology by providing financial incentives to go directly to natural gas or better, to alternative power. Those in the environmental movements might consider this cause as one of the better routes to a brighter future for the planet, a viable alternative to trying to save forests in undeveloped areas more directly.

The list above is intended to serve only as a starter for discussion. Clearly, in a few years the details of some of the rapid changes in society and technology predicted will become clearer, as will the increasing evidence and implications of climate change. The economics of carbon reduction will become clearer. Promising, relevant new technologies will almost certainly emerge. The climate changes to come by 2035 are relatively slow-moving compared to the present cultural changes and largely inevitable by now. Climate change is a relatively long-term problem, and it will take a number of well thought-out, long-term solutions, solutions that will remain valid even while our global society goes through rapid, uncertain but incredible changes. It seems very likely that the situation and any additional necessary actions will be much clearer when reevaluated in 2035.

Coping by Individuals and Families

Given that global climate change is already well underway and many further changes are inevitable, the question of adapting to climate change should also be addressed on the personal level. Global warming will not affect all areas equally or at the same time. Of course, some outcomes, like higher food prices, could affect everyone on the planet. But climate is always expressed in weather, and weather is local and personal. There are considerable differences in what we can expect for changes in different areas due to global warming.

For example, I live in San Diego. Past experience, including pre-historical records, our most recent weather patterns and predictive models all suggest that the main climate change that will directly affect this area will be a continuation and increase of drought in the American Southwest. A decline in water available for urban and agricultural use seems probable. However, it seems unlikely that those of us depending on public water in urban areas will need to move for this reason alone, at least not without some warning. Agriculture uses about 70 percent of all fresh water, and recent experience has shown that conservation can easily reduce the amount used with only modest price increases for foods (Climate Central, 2011). There is plenty of room for improvement globally, less than 4 per cent of irrigated water uses drip or other micro irrigation (Worldwatch Institute, 2013). Much of the portion of water going to cities in developed countries could also be saved, in that it spent on things that are optional, like expanses of green lawns, golf courses, etc.

My experience living through the Marin county water shortage (1977) is that a concerted public effort to save water means that an urban area can live on a fraction of the water normally used. If water gets really short, we can always give up the idea that we have to be able to drink the same water we use for showers, toilets, lawns and gardens. Changing building codes to allow people to more easily install grey water and rain water systems will be helpful as well.

The expected rise in sea level will, of course, affect the parts of San Diego that are very close to sea level. Checking the NOAA web site, however, I find that the sea level here is currently rising at the net rate of 2.06 mm/year. This is only 8 inches per century, less than the average global rise in sea level because land on the West coast is also rising. If I lived at sea level here - as many do - I would keep aeye on developments and plan my life so that I would be able to move to higher ground if and when needed. One reason for keeping a close eye on things is that the most recent measurements suggest that the average rate of sea level rise is *accelerating* due to increased melting in Greenland and Antarctica. Current predictions of *future* global sea level rises have doubled, to a total of one foot by 2050.[58] More worrisome, Hansen (2009) suggests that IPCC estimates of ice melting from model calculations have been seriously underestimated, and seas could eventually begin rising at the rate of 3-5 meters/century. His fears are based on the maximum rates seen during the last major global warming event.[59] This would diminish the time required for a response, to put it mildly.

The suggestion to most people on the personal level is essentially to continue to live your life as you are, but to learn about what is in store for you and your area, keep an eye on changes in the predictions as time goes on, as well as on the scientist's evaluations of the extent to which the changes are already occurring, and to live your life so that you are psychologically, financially and physically able to cope with changes when and if they do occur. For many people, given the unpredictability of both our society and the climate, it would be wise to keep a certain flexibility in where you need to live and work, for example, and be able to move either with relatively short notice. Being willing and able to make appropriate changes will be particularly important for those involved in food production or other activities critically dependent on local weather.

8. EDUCATION IN AN ERA OF CONSTANT CHANGE

Processes of learning and education will need to change to allow children and adults to make the best use of future opportunities, especially opportunities offered in a future in which change is constant. Potential contributions to the processes of learning have already begun to evolve out of the very processes driving the high rate of change, and this tendency will only increase in the future.

Content and Goals of Education

In this section, we look at changes required in our educational system in order to adapt to an era of increasingly rapid rates of change. Relevant to this issue is the fact that, to many educators, it is clear that our education has not even been keeping up to changing demands of the recent past. I will comment here specifically only on subjects that I have taught, leaving criticism of other subjects to teachers in those areas.

The first thing I would like to ask is: Why do our children still have to learn how to multiply and divide multi-digit numbers? I have been a practicing scientist in some of the most math-intensive fields of science, and I know that I would never multiply out two multi-digit numbers by hand, not since the pocket calculator became available. Or divide two such numbers using paper and pencil. Doing so makes little sense today when a calculator app is available on every mobile phone, and even most children seem to have one of those with them at *all* times these days.

It does not seem to make sense these days to take the time to require children to learn to do a number of different arithmetic skills by hand. It would be more effective to provide them with a calculator, teach them how to set up problems and then how to use the calculator effectively for the calculations. After all, unlike talking, learning arithmetic skills like long division has never been an evolutionarily favored skill for humans. It is not "natural," and in fact some fraction of people (at least 6 percent) have an organic brain disability, called dyscalculia, that completely prevents them from doing so.[60] This condition is estimated to be just as common as the reading disability, dyslexia, but much less well known and often unrecognized. People afflicted with dyscalculia cannot memorize even simple multiplication tables, and the only reasonable solution is to allow them to use calculators. Why not let everyone do so?

An important example of the failure to update requirements to meet reality was revealed when tutoring adult clients who were currently in the military and needed to pass a test required to move up to the next pay grade. The ability to do typical grade school arithmetic was a significant fraction of the test, and calculators were *not* allowed. My adult students, who I should add had obviously passed the military entrance exam and had been performing their assigned duties quite well for several years, had severe dyscalculia and were simply unable to learn to do arithmetic fast enough to get a good score on that section. Think of that: we are denying people advancement in their chosen profession by requiring them to do something that *no one* in their profession or society actually has to do anymore.

Even most normal American adults, who were supposed to have been taught fraction manipulations as children, no longer know how to "do fractions" or even why. Teaching people skills that they will never need as adults is not helpful. In contrast, teaching people how to think quantitatively about a problem, how to properly set up the relevant calculations, and then how to use a calculator or even a voice-operated computer to carry out the calculations *is* likely to be useful. Teaching people to learn to understand scientific results, to understand simple statistics, to learn to interpret data and to understand graphs, etc., is also likely to continue to be useful for applications in the home and workplace, and to be especially appropriate for informed citizenry in the modern society.

Traditional math instruction focused on computational methods is not even useful. Research by Jo Boaler (2008), for example, found that many people forced to learn math through traditional computational methods report as adults that they "hate math." Even when these adults successfully do use math in work or everyday life, they almost never use the computational methods they were taught. Boaler cites as an example a cook who had to take 2/3 of 3/4 of a cup of sugar. Rather than multiplying the fraction as she had been taught in school (to get 1/2 cup) she measured out 2/3 of a cup, poured and flattened it, and removed a quarter. In contrast, Boaler reports that people who have been taught math by a problem-solving approach report that they "like math" and they do use methods they learned in school in their work and everyday life.

My main point is that it seems to take a long time to change the content of our school's curriculum to reflect skills that actually are needed in the current society. This is perhaps more true in math than in any other area, and this is a especially important problem because quantitative reasoning is expected to become even more important in the future. For example, a journalist recently noted with horror that 42% of Indian engineering graduates did not know how to divide two decimal numbers by hand.[61] This was discovered on a test supposedly evaluating them for skills needed for employment. What kind of company would ever require their engineers to do such a calculation by hand? What kind of thinking leads an educator to assume that this ability should be tested for *and* that the absence of this ability is a tragedy? Why weren't they allowed to use calculators?

This brings up the central question: Why are schools (and society) so slow at changing the curriculum content to match skills actually needed? If they lag so far behind reality now, how much worse will it be when the change rate increases as much as it is expected to, when they have to prepare students for a rapidly changing society? When it is not even clear to *anyone* what skills and content will be needed when they leave school?

The resistance to changing the formal school curriculum is reflected in social norms, and the origin in both places seems largely to be tradition. Worse, curricula in schools are typically set by committees, and everyone on the committee has their favorite subject which just must be included, and, in the end, it seems that most committees operate by giving in to everyone. And everyone seems to feel that, if they had to learn it, everyone should.

When teaching a high school physics course, for example, I was amazed by the huge list of topics that California state standards required to be covered. Learning all these detailed applications of Newtonian physics seemed to leave no time for students to learn to appreciate the process of science. Understanding how to determine and answer those questions which *can* be addressed by science seems by far the most important outcome of any high school exposure to science.

One can therefore argue that our schools are not now preparing students well for the kind of society we *already* have. If so, it will be even more difficult to change our school curriculum to prepare students for a society that we don't yet have, and which is certainly uncertain in its details.

Changes in the Process of Education

Fortunately, the *process* of education seems a bit more amenable to change than the content, even for changes involving traditional school systems. These changes are somewhat easier to implement because they are, in part, market driven, i.e., driven by the need to reduce the costs of education. We will also see that many needed changes are likely to be provided by other educational institutions and processes that can (and will) change faster than the formal educational system can.

The most recent major push in formal education is to reverse the present system where the teacher lectures during the day, and students do homework at night, to one in which students learn by computer at school or at night and the teacher helps them do the homework during the day. The computer learning may involve simply watching lectures from an expert teacher available over the internet to learning content using individually paced instruction. Adaptive programs can track student learning by asking questions or giving them typical problems to solve, then modify the teaching approach as needed by the individual student. There are many companies currently competing to bring their software solutions to the K-12 classroom.[62]

The massive open online course (MOOC) is achieving wide acceptance as a method of providing college level education to people throughout the world, people who may not have the opportunity to take a quality course otherwise. I note that people locally are using the web site Meetup to organize study groups for students who sign up for such courses, so that people can help each other with homework - people who would otherwise not know of each other's existence.

If you believe the predictions for the immediate future (Figure B3 and Table 2), things are going to be changing at an ever faster rate. Further, while we can predict some probable outcomes, we don't even know which technology will provide the most significant changes ten or more years down the road. And, even if we think we can identify the technology, the applications that will emerge and the consequences for society are largely unknown. And the consequences are *unknowable,* as our recent history with the internet shows clearly.

People born into an industrial age society grew up in a culture with a change rate having a doubling time of about 50 years (in personal income). That means that their formal education *could be* (wasn't always) planned with a reasonable expectation of what the future would be like for them. Subjects and skills were taught that were expected to be useful for their future lives, with reasonable success. As an engineering student in the early 60s, for example, I was taught how to use a slide rule. Calculators at that time were large desktop machines that could do simple arithmetic but not much more. Pocket calculators did not replace slide rules for engineers until the 80s, and they were easy enough to learn to use.

In contrast, if you assume the very speculative projections in Table 2 are anywhere near the mark, children born today and in the next few decades will grow up in a society with a change rate having a doubling time of 18 years or less. *Major aspects of their society will change completely in the time between the start of first grade and college graduation.* Any content that students learn in school may be outdated before they graduate. Skills and attitudes learned in school may not be useful when they begin their post-graduate existence. This will be an even more severe problem for adults no longer in the formal educations system, unless special programs are adopted.

One thing will remain the same: the high rate of change will be constant. What people will have to learn are the skills needed to cope with a society with a persistently high rate of change. They will need to learn the skills and the attitude that make each new development an intriguing challenge *and* one that they feel they have, or can learn, the skills to meet. They will need to be able to continue learning, as needed, throughout life. They will need, in other words, whatever it takes to be happy and productive in a society with a high change rate – in fact it would be best if every individual could learn how tomaximize the *advantages* of living in such a society.

What is required to meet this challenge is an attitude toward education and learning processes that match the forces that are producing the huge changes themselves. That is, we will need to make room for people and societies to allow the learning marketplace to flourish, so that *the same forces that are driving change become available for learning how to cope with the change.* I expect that innovations in education will largely become available as the need does, from all the same processes - enabled by computers and the internet and whatever follows. I do not know the details, of course, none of us can know the details, of what technologies will emerge to meet our needs. What we will need to do is to structure our education and our educational systems so that they allow opportunities for useful educational systems to evolve, be developed and accepted. These solutions will allow each of us in our own learning to take advantage of emerging technologies in ways that may be more or less unique to each person. Businesses, governments and military will increasingly need to be able to give credit for talents and learning obtained in the non-traditional channels that will become available.

In sum: the goal of education in the broad sense should be to give people the inner resources to allow them to cope with and even *benefit from* living in a society with a high change rate, and the ability to generate, locate and use external sources of aid, sources provided by the society whether by human, robot, computer, internet-enabled means or whatever else may emerge.

9. WHEN AUTOMATION AND ROBOTS TAKE OUR JOBS

Most traditional jobs now held by humans will be taken over by robots. This will create opportunities where humans, freed of routine work, may find new and innovative ways to spend their time and energy. The human-internet connection will continue to be a key source for work and other activities, providing solutions we cannot foresee to problems that we don't expect. Questions about "where money will come from" in the digital society to come are discussed, in part by using an illustration of money flow in a simple shantytown economy.

Many Traditional Jobs Will be Gone

Digital technology has consistently increased the usefulness and areas of application of industrial robots (which include agricultural robots). There are presently a total of 1.4 million industrial robots installed worldwide, and installations are increasing at a net rate of 5.4 percent per year.[63] If growth continues at this rate, by the year 2040 there will be 9 million industrial robots installed, one for each thousand people worldwide. In the U.S., there will be one industrial robot for each *fifty* workers by then (see Figure B4).

Futurists who address this issue predict that most traditional jobs now held by humans will eventually be replaced by robots, and that the intelligence and capability of computers and robots are almost certain to continue to increase exponentially (e.g., Ford 2009, Gore 2013, Mulhall 2002). Note that industrial robots are large machines which at present cost on average $150,000 each and which, at this cost, *must* displace human workers when installed. Because the machine necessarily costs less than the worker it replaces (per unit of production), productivity increases when jobs are automated. ***Therefore, GWP and GWP/person worldwide will automatically increase as automation does, which is the good news.***

The number of unemployed workers may also increase, however, which is potentially the "bad news." At least, there are relatively few ideas currently being discussed about what to "do with" workers displaced by automation. *They don't necessarily need to remain unemployed* (or at least unoccupied, as discussed below.) Studies by the International Federation of Robotics (IFR) claim that, at least in developed countries, each industrial robot installed creates 3-5 new jobs, most of them "downstream" of the installation. Unfortunately, one contributing factor to this job creation – and the IFR doesn't say what fraction – originates in the fact that labor costs in most undeveloped economies have been until recently too low to allow automation. As a consequence, much manufacturing requiring cheap labor had previously moved "offshore" from developed nations. When developed countries successfully compete against countries with low labor costs by automating production, *and* when wages rise in undeveloped areas, prices of production in developed areas become competitive, at least locally, and jobs move back "onshore" (see Meisel, 2013). This fact can account for at least some of the "job creation" effect of automation. That is, some of the jobs counted by the IFR may be not so much *created* as being *restored*, or "reshored" (Cowen, 2013).

At this point we can be absolutely certain only that worldwide automation will, at the very least, reduce the number of boring, repetitive factory jobs. It is actually not obvious how worldwide automation will change the total number of jobs held by humans, at least in the immediate future. However, those displaced by automation or other use of robots will at least have to change to different jobs, or find other things to do with their time.

Many New Job Opportunities Will Appear

At least at first, people displaced by automation may find or create jobs doing those kinds of things for people that computers are not very good at: creating art and music, packaging and selling services and products; in short, the many possible creative activities in which human taste and enjoyment are involved (Reese, 2013). For example, Apple products have consistently demonstrated that human innovation in design and product interface remains a very important and desirable aspect of digital product development.

Brynjolfsson and McAfee (2011) and Meisel (2013) also suggest that some displaced workers can usefully become involved with the type of jobs which pair people with computers to do work that neither can do well alone, e.g., checking computer-generated verbatim transcripts. Some of these jobs are available online, so that anyone with a computer and internet connection can find paid work on a job as a "contributor." Such "crowdsourcing" businesses currently include Mechanical Turk and CrowdFlower. While the pay is typically low by U.S. standards, there are currently estimated to be a total of a million contributors worldwide working on these two sites.[64] Finally, Mulhall (2002) suggests that governments sponsor self-help groups for people displaced by automation. A decade later, I think that it has become obvious that if this is going to needed, and it is, it will happen (already is happening) naturally via the internet; and government action will not be needed, except to get out of the way.

In contrast, Randers (2008) simply assumed that government action would be taken to maintain "full employment" to avoid social unrest. It seems likely that his assumption was mostly for convenience in his model calculations, and otherwise I don't think this is likely or even desirable. If likely, why has the U.S. government allowed unemployment to hover around the historically high (and politically dangerous) level of ten percent after the last downturn, as companies continue to invest in automation rather than immediately rehire employees?[65] The trend in automation is as clear as it is unstoppable. It will only increase, and eventually robots will supply all of our traditional needs for manufacturing, as well as for many services. It does not seem at all desirable to leave it to governments to decide what we are going to do with our time when automation displaces us. Some governments might force its citizens to keep occupied with a "make work" job and I think none of us want to be forced to be a Moscow-style sidewalk sweeper to qualify for financial or other assistance.

To best take advantage of future opportunities, Brynjolfsson and McAffee (2011) suggest that our educational systems should be restructured with a focus on how to invent new processes for machine/human interaction, along with vastly increased training in the entrepreneurship skills needed to implement these ideas in the world. It is likely that many such educational opportunities will be (have already been) invented naturally through, and enabled by, the internet.

The development of the internet has enabled (will continue to enable) many new opportunities for people to make money and/or have something enjoyable to do with their time, and more are being invented all the time. The most obvious examples involve the development of a wide variety of businesses, businesses that use the internet but do not require a brick and mortar store. These "e-businesses," offer an increasingly popular way for individuals to find useful, interesting and potentially remunerative activities. There is no lack of advice on how anyone can take something they enjoy doing at home and turn it into a money-making business, without ever leaving home. Searching "online business" on the Amazon.com bookstore brings up over a hundred thousand titles, many of them "how-to" books (e.g., Ostrofsky, 2011). The growth in on-line business is currently responsible for a large fraction of the growth in GWP each year, particularly in less developed regions.[66]

The different kinds of on-line businesses enabled by the internet are essentially unlimited. A simple example is my recent experience using the internet to build a professional service business. After I retired from the university, I decided that I wanted to start a math tutoring business. I developed several web sites focusing on different ages and different kinds of problems learning math. I placed targeted ads on Google, to run only when local people searched for math tutoring, and the ads sent them to one of my sites if they clicked on them. By careful choice of key words, often I got placement in the top spot on their search results page. The ads were reasonably inexpensive because I had to pay Google only when people clicked on one of my ads. A satisfactory percentage of clicks resulted in calls from prospective customers. Fairly quickly, and with much less expense and effort than would have been involved in pre-internet times, I began to build a reasonable business, bringing in students of all ages.

In contrast, I can attest from personal experience what it was like in the pre-internet days to build a similar business. In the 1980s, I went into private practice as a psychologist. It took several years to build up a practice and it was a lot of work. I didn't make rent for a while, much less an adequate income. Back then, there was no similar way to announce your practice via the internet, no way to build it except the very slow and random word of mouth. Now you can announce your business online, have customers post reviews, and have a presence on a variety of online platforms.

Other new opportunities for remunerative activities are being invented every day on the internet. For example, part of Amazon.com's success is that they make available a very easy-to-use process for anyone to sell products through their site. Through them, as a minor example, I recently sold a large number of used books, many of them old textbooks, that I had accumulated over many years of working in different fields. I made some money and had the satisfaction of seeing all of my old books go to good homes. These were all books that I had previously carried around to used book stores in big boxes, or tried to unload at garage sales, completely without success. The new, internet-enabled solution was *much* better than any previously available solutions.

Amazon's services also allow some people to earn a reasonable income in the business of "retail flipping" buying toys, consumer electronics, books, etc., on sale at retail stores, and selling them at higher prices on Amazon.[67] There are similar opportunities, each with their own particular strengths and weaknesses, using the business models of other services including Craigslist, eBay, Facebook and Apple (e.g., Hawkins, 2012).

In general, Marc Ostrofsky (2011) offers a highly recommended book with a wealth of ideas about how to make money in any aspect of the internet. There are also many books on more specialized topics, such as how to start a blog or website that people will read and that makes you money (Omar, 2013) or how to self-publish a book to sell on Amazon (McMullen, 2012, 2013). There no longer seems to be a stigma to self-publishing over 70 percent of books are now self-published, including many by well-known authors. By self-publishing, authors have the chance to control the book design completely, publicize the book themselves, and keep all profits. This book is, of course, an example of this process. A largely unrecognized advantage of publishing using a "print on demand" publisher like CreateSpace is that the upfront costs are *zero* and one can update or improve a publication at any time without difficulty or any extra cost (as this book was).

Opportunities for on-line businesses are increasing rapidly because more and more people are now searching the internet first when they want to buy something. The number is growing rapidly. Internet sales increased 16 percent during last years' holiday shopping season in the U.S. compared to the year before, and accounted for a total of $39 billion in sales.[68] While total on-line sales only accounted for about 5.4 percent of all retail sales last year, the growth in on-line sales was the only bright spot in an otherwise dismal year for retailers.[69]

An Increasing Number of People Will Not Want or Need Regular Jobs

Of course, the internet offers *much* more than new opportunities to make money. Most of the sites mentioned above owe their existence not from making money - that is an add-on - but by providing sources for entertainment, news, information, games, gossip, contact with friends, etc. More people are now using Facebook to communicate with friends than email, for example (Ostrofsky, 2011). Many people frequently share their favorite pictures or videos by posting on YouTube, without any expectation of monetary return, being pleased to share, perhaps hoping for the thrill of having a video go "viral." Of course, having a video go viral *can* lead to business opportunities later.

For more serious pursuits, more and more people are finding that the only practical way to meet possible mates is through sites such as Match.com, eHarmony.com, ChristianMingle.com, etc. This seems to be particularly true for older adults who do not have much opportunity to meet possible match-ups in their normal lives. A major advantage of dating using these services is that you can be sure you are meeting people who have the same objective as you do, e.g., a long-term relationship. And the photos on their site give you some idea of what the person looks like (at least looked like once). A standard bit of useful advice for such activities is to meet people you don't know at a public location, and check them out using online resources before going further.

The internet will offer more and more opportunities for people to interact with others, most of which have not yet been imagined. Some of the services that the internet will offer come from areas to be discussed below, i.e., providing services that help people whose previous jobs have been displaced by automation.

For the majority of those displaced by automation, I suggest that the solution may lie in any of a wide range of activities, most of which involve using the internet: social connection, education, self-expression, specialty businesses of all kinds, political activity, charity work, and a *huge* number of other activities that have not even been dreamed up yet. In other words, many people may become happily "employed" in activities which do not look at all like traditional employment, and which may look more like (very early) retirement. *Overall, the "bad news" of automation could turn out to be one of the* *greatest boons to the human race*. It certainly is a great opportunity, and the best kind of great opportunity, one we cannot avoid.

In general, I expect that digital technology itself will (continue to) provide answers to many of the problems it creates. In particular, *job destruction due to digital technology will be counterbalanced by job creation due to digital technology* as well as making a wide variety of attractive alternatives to jobs available.

As another example of a digital solution to a problem caused by digital technology consider the increase in traffic accidents which has arisen from people texting while driving, or trying to look up directions on their smart phone, etc. Even hands-free mobile phone operation has been found to result in increasing accidents by distracting drivers. When I early on predicted this, people would object that it shouldn't, because people talk to other people in their cars. But the people on the other end of *that* conversation are in the cars too and if the driver fails to notice something important, they will point it out to them ("Watch out!"). It turns out that even having additional people in the car doesn't help if all of the people in the car are teenagers. In this case, the accident rate goes up as well, because the other teenagers apparently do not watch the road as nervously as more experienced drivers do.

The solution provided by the digital age: having computers drive our cars for us. It has already been found that technology like radar braking, automatic braking systems, and non-skid or rollover control significantly reduces accident rates. In the end, we will end up with full digital control of driving, and a near-zero accident rate. And we can be online as much as we want.

Most people who look at the question of the future of work for the human race completely miss the possibility that there is anything else that humans might choose to do with their lives, if they didn't have to work. The alternatives suggested above should be compared to Tyler Cowen's arguments[70] that those who will prosper in the immediate future will be: the conscientious, people who listen to computers, people with a marketing touch, motivators, people with delicate feelings, political radicals, and *people who don't need money*. The first five categories are all discussed to some extent above, and involve more or less traditional jobs where new interactions with computers and the internet will be primary. I am suggesting, however, more: that an increasing number of people will eventually end up in the last group, a group characterized by Cowen as "people who are bright, culturally literate, Internet-savvy, and far from committed to the idea of hard work directed toward earning a good middle-class living." Further, while this group will technically be lower class, they will not *feel* poor, and they won't *be* poor, if their standard of living is compared not to their peers, but to people in the past. Instead, they will feel glad that they are not in the rat race, that they are having too much fun to work at a regular job.

Where will People Get Money Without "Regular" Jobs?

This question concerns many people, especially those who have had traditional jobs most of their lives, people who were forced to work because they lived in an economy of scarcity rather than abundance. To best answer this question, it is first important to understand where money comes from, what it stands for, and how wealth is generated in a society.

Matt Ridley (2010) provides persuasive arguments, and examples through the ages, to show how wealth is generated by trade, and from the specialization that trading allows. If I am growing corn, and need shoes, it is more efficient for me to grow a little extra corn and to trade for some shoes from someone who specializes in making shoes but needs corn. *The more trading that occurs, the more wealth is created.* This is NOT a zero-sum game. In our modern society, of course, money acts as a marker to facilitate trade so that I can sell my corn to a grocer, and don't have to count on my shoemaker needing corn at the same time as I need shoes.

The "shantytown" example. The whole concept became blindingly clear to me while visiting a new neighborhood formed by squatters who arrived last year to become part of the huge Villa El Salvador settlement in Peru. The refugee group that was occupying this government land had to live on it for ten years to earn title, at which time they would be entitled to schools, water, sewers, electrical power and all the other perks of official residency. In the meantime they had self-organized to live as best as they could. Each family had taken a twenty-foot square to build a house, using whatever materials they could find. The houses were organized in blocks, with streets in a grid, all wide enough for one car to pass, but of course nearly everyone walked. In each block, I noticed that one or more houses had opened small businesses, in the front half of the house or on the first floor of two story residences. There were small grocery stores, fruit stands, newspaper kiosks, clothing stores, CD stores, internet kiosks, cell phone stores, even beauty salons. It was the beauty salons that made the most impression. Where did these obviously poor people get the money to afford to go to a beauty salon?

It was the more puzzling because it was obvious that the customers for all these neighborhood businesses could come only from this area. The neighborhood was quite isolated and there were few cars in sight: these businesses were not bringing in any customers or money from outside. All the businesses were in residents' homes and run by residents. They were providing an obviously needed service to their neighbors, as there were no regular retail outlets in miles.[71] But where did the money come from to support all these small businesses?

I was told that most of the men were able to find occasional work in nearby Lima, mostly as unskilled day laborers. Some women did too, but most worked within the community. Those finding outside employment brought money back to the community. Edwardo[72] might give ten dollars to his wife, who might spend it to buy food from the local grocer. The grocer might use some of the money to buy clothes. The clothing seller might buy her daughter a CD. The CD seller might use the money to buy mobile phone time so she could order more CDs. The mobile phone dealer might use the money to get her hair done. The beautician might contribute money to support the teachers in the children's school in the settlement. The money eventually might get back to Edwardo's wife, when she earns it cooking in the community kitchen. She would give some to Edwardo so he would be able to take the bus back to the city and earn more.

Most of the money seen in this community is created by circulation within the community. Each time it changes hands, of course, some of the money *eventually* has to go outside of the community to buy materials the community does not generate themselves, such as food stock, mobile phones, beautician supplies, etc., but an equal amount of money comes back into the community by the men who exchange their labor for it outside the community. Figure 3 illustrates this process.

The outermost loop represents those who find work outside the community, and bring money back into the community. This money circulates in the community as each person buys and sells goods or services to their neighbors. The next loop inside represents the money that has to (eventually)[73] go back out to wholesale distributors to buy food and other goods for retail sales, hair dressing supplies, etc. While this does reduce the amount of money circulating each time it changes hands, especially for those selling goods instead of services, estimates in general suggest that local circulation produces *several times* the amount of money coming from outside. For example, when a new company moves into an area, the effect on the local economy is usually determined by estimating the number of additional people who are provided jobs for each person who is hired directly by the company typical estimates are 3-6 workers. Such money generation is a well-known part of economics: the Villa San Salvador group is striking only because it is easier to identify the probable money flows.

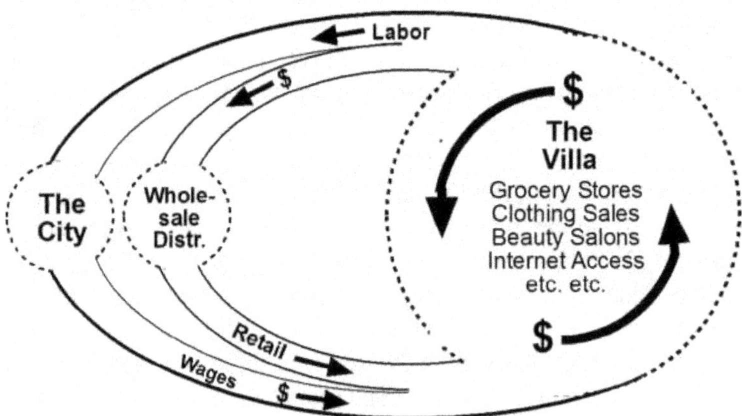

Figure 3. Where Money Comes From, Villa San Salvador

The same idea applies to the much larger, global emerging digital society, sketched in Figure 4. The diagram is divided into two groups: All human to human interactions on the right, and, on the left, the production of goods and services from industries that are largely automated. The outermost loop is meant to suggest that, for a time at least, some people will have high paying and/or creative jobs in (mostly) automated manufacturing or services because they have skills that robots and intelligent machines still don't have, or we don't want to give them planning, management, product design. These professionals will, in general, multiply their talents by working cooperatively with the different talents of intelligent machines.

The next loop inside indicates one of the main ways people will earn money in this age: by return from their capitol investments, in particular by holding stock in companies that use robots to efficiently produce everything needed at a very low price (Cowen, 2013). Of the money that comes back to people (lowest loop) some of it is used to purchase goods and services from the automated system, shown by the innermost upper loop. The inner return loop in this case, labeled "goods & services," is shown larger than the cost loop, to indicate how much more you can get for your money with automated production. Automation is a multiplier of money in this sense. Note that, as drawn, there is more money coming into the human group than going out, so that the human group becomes richer with time and needs less and less to work for money.

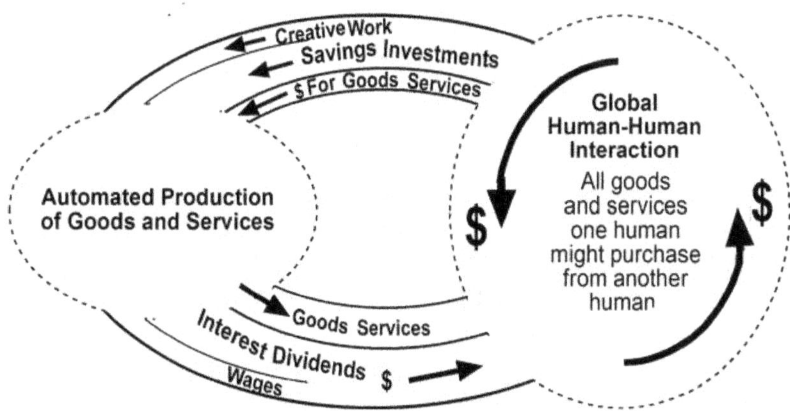

Figure 4. Where Money Comes From, Digital Age

The money that these income streams bring back to the global human community can circulate within it and be amplified in the same way as in the Villa, when people exchange their products or talents with others. Most people won't have to work, of course, because the cost of living will be so low. They would be able to get enough money to live from the return from even modest capital assets. Those of us who choose to work to get money will not have to work very much, unless we want to, and we will largely be able to do work we enjoy as many of us do already.

No one will want to dig ditches or pick up garbage, the robots will do it. But there will be plenty of opportunities for people who want to help other humans look more attractive, or learn how to do yoga, or walk their dogs or robodogs, or make art or music, or write books or blogs, or design clothing or other articles for human use, or invent games and toys. In addition to all the other jobs that people will be able to do that haven't even been imagined yet, in other words, there are all the jobs helping other people (learn how to) be (more) human, to enjoy life more, to live creatively, to best interact with all aspects of their society, human and machine. As Cowen (2013) notes, there will be many jobs available working for the richest groups, coming up with innovative ways for them to spend their money.

Finally, it should be pointed out, that while automation may replace nearly all of the routine production and service work that humans now do, it is still not possible (and may not be easy or cheap for some time) to make a robot that is as good as a human is *as a total package*. We animals have millions of years of evolution behind our design. We are still having trouble making humanoid robots who can walk across rough terrain as well as a human can, who can move as quickly as humans do, with adequate strength, who have as good a control over movement and touch, who are as smart as humans, who can use a variety of materials as an energy source, and do all of these things at once. In human society, at least, we may find out that it isn't so easy or desirable to replace all human activity with humanoid robots. A robot which can do all the things a human can do (and as well) may not be possible for a long time, and even when possible, may be more expensive in direct costs or environmental damage than simply paying a human to do the same job. Unless, of course, the job is in space or on Mars where robots clearly have an advantage over humans.[74]

When you wonder what you will do with your time, remember that technological progress through the ages has already eliminated a lot of the most boring human jobs, especially for people in developed societies. Do you know anyone who cuts down wheat with a scythe? Or carries "honey buckets?" Do *you* miss having those jobs available? We still seem to be able to keep busy in spite of many such jobs already being automated and no longer available. In spite of having to work less in the present era as a result of such past changes, the essence of being human seems to be that we still seem to find *plenty* to do with our time.

A demonstration of the reality of wealth creation for the classroom, etc. Make up hats or name tags with letters on them (A,

B, C..) and a set of cards or small boxes with the same sequence. Give everyone a hat and a box with a different letter. (Make sure that, overall, the set of hats and boxes match.) Tell everyone that they will sell their box to anyone who offers them ten dollars for it. What each of them wants, what will most improve their life, is to get a box which matches their hat. The box could represent whatever thing or service they most wanted right now. Say, "I am the government, I just printed a ten dollar bill, and I am loaning it to one person," and give it to, say, A. The person A is then to find the person with the A box and buy it from them, whereupon that person will take the ten dollars to buy their box, etc. If A gets back the $10 before everyone is done, have him loan it to a person who hasn't yet got their box, and keep doing this. When everyone has got their box, note to the group that everyone got and spent lots of money, but there was only $10 in circulation and no one had even earned that money. Money is no more than a marker to facilitate trade, and, as such, there is essentially an unlimited amount of it. But make sure you get your ten dollars back from the last trader.

What About "Job Satisfaction?"

People who have (have had) fairly good jobs, people who are (were) professionals, often ask at this point: "Well, ok, maybe we won't need jobs for money but what about all the other perks of having a job, the satisfaction of accomplishing something useful in the world, the social interaction that goes with a job in an office, and all the rest?" I say what about the joys of rush hour traffic, having to run an important meeting with a bad cold, having bosses that pass stress down hill or subordinates who passive aggressively reflect stress upward? I say that there is nothing that a job could give you that some other activity or combination of activities might. Are there still people in the world who need some help? Start a charity to help them, and succeed in being one of the few charities to actually give them the benefit of most of what you collect. Or help directly through volunteer work. The kinds of social/business interactions in the world that you could get involved with are essentially infinite, and with the internet, you can connect to people with the same goals or interests that you have no matter where you live.

And the alleged satisfaction from having a job? How we feel about working is determined by our cultures' attitude toward work. If our society needs us to work, it will value this and we will feel valued and in turn value our contribution. If not, it will lead us to value other things: creative skills, people skills, volunteer and committee work, whatever one does to be a useful member of a community will be valued.

When there is this kind of transition from a society that values traditional work, particularly when it is as fast as expected, the values of adults will lag those needed by the culture as it changes, and, of course, these adults will feel some stress or discomfort as the attitudes they were taught as children are no longer as useful. Aha, more job opportunities! Psychologists, counselors and others will find useful occupations helping people having difficulty making the transition. Isn't it obvious, that if you want help adapting to a new society, you might prefer to consult with another human being who has faced and dealt with similar issues, rather than a robodoc shrink who - no matter how competent - even could be seen as a symbolic part of the problem?

As for people who don't want to change: The culture of this future society is expected to allow even more range for personal choices than allowed in developed economies today, so those who don't want to change their values e.g., those who really want to continue working will largely be able to find ways to do so. But, fewer and fewer people are going to *have* to work as automation takes over.

Handled well, the advent of automation will turn out to be one of the best things that ever happened to the human race, freeing all of us to do whatever we want to do with our lives, and allowing our culture to continue to expand exponentially.

10. MACHINE SUPERINTELLIGENCE AND THE CHALLENGE TO THE HUMAN RACE

> Once computers exceed human-level intelligence, it is considered likely that artificial superintelligence (ASI) will soon ensue, when computers become able to design computers who are more intelligent than they are, bootstrapping machine intelligence to very high levels. This implies a *singularity* will be reached past which our limited human intelligence is simply unable to see. While this certainly represents a challenge to the human species, it also seems the most obvious direction for the evolution of our civilization, in which our civilization spreads out to the stars.

Looking at the Coming Singularity

If the arguments in the previous chapter are correct, the fact that machines will soon take over most of our current jobs does not mean the end of the world for the human race. Rather, this stage may indeed constitute yet another opportunity (in a *long* list of similar steps through the ages) for us to free ourselves from drudgery, and to find more exciting things to do with our lives.

The advent of machine superintelligence seems more likely to lead to problems for the human race, however. The question is: what will happen if and when machines become intelligent enough to direct their own evolution? This is not a new concern. In 1965, mathematician Irving J. Good wrote,

> Let an ultraintelligent machine be defined as a machine that can far surpass all the intellectual activities of any man however clever. Since the design of machines is one of these intellectual activities, an ultraintelligent machine could design even better machines; there would then unquestionably be an 'intelligence explosion,' and the intelligence of man would be left far behind. Thus the first ultraintelligent machine is the last invention that man need ever make.[75]

A specific prediction was made in 1993 by Vernor Vinge, at the time a professor of computer science at San Diego State University, who wrote, "Within thirty years, we will have the technological means to create superhuman intelligence. Shortly after, *the human era will be ended*" (Italics added).[76]

This time scale appears now to have been a bit optimistic (or pessimistic, depending on your view). To reach superintelligence, computers will first have to reach the level of artificial general intelligence (AGI). Estimates made in Chapter 4 suggest that, by about 2035, AGI will have developed enough to make a significant impact on humankind. Some experts (see Barrat, 2013) suggest that superintelligent machines may have already arrived by then, or may follow quite quickly thereafter. Speed of development will be increased exponentially if and when intelligent machines reach the stage where they can design machines that are even more intelligent, quickly bootstrapping machines to *very* high levels of intelligence. With speeds of digital processing doubling every year (Table B1) hardware limitations are not likely to prove a barrier to this kind of development for very long.

Unlike other problems currently facing humanity, the outcome of this one is uniquely difficult to predict. We have no previous experience with beings who are smarter than we are, certainly not of beings who can interact autonomously in a technological world. We don't have any good way to imagine what it would be like to interact with such beings, whether biological or machine derived. We really have no idea what they could be like, and what the range of possibilities ultimately are. Nothing like this issue illustrates more clearly the impossibility of being able to predict the *consequences* of a particular technological advance, while nonetheless being able to predict that a certain technological advance probably *will* occur.

Verner Vinge recognized early the impossibility of predicting the future once machine superintelligence was reached. He labeled this point the *singularity*, in analogy to the singularity in space-time that defines a black hole. When gravity becomes large enough to form a black hole, it warps space enough that light can't escape: one cannot see into the bottom of a black hole, can never know what happens there. Once superintelligent machines are developed, and they begin to design machines that are even more intelligent, there will be an "intelligence explosion" and the mere human intellect will not see the outcome past a certain point. The future becomes a singularity to us at this point, in that we cannot see beyond it. Note that this definition is consistent with its use in physics, mathematics and logic; it is therefore the definition adopted in this book, and as such it is slightly different from the definition adopted by some others (e.g., Kurzweil, 1990; and Mulhall, 2002).

What Does the Immediate Future Hold?

In the immediate future (say, to 2050) it seems unlikely that computers will be able "take over," in the sense of beginning to autonomously act against biological humans' best interests. Unlike us, computers require technology just to survive, let alone to reproduce, i.e., they need technology to produce and support new computers or robots. I think, for the near future, we humans will be able to control the source of components and subsystems. We will know where the "plug" is, and be able to pull it if we need to. Of course, as we become more dependent on computers, as we already are, pulling the plug completely could be a serious problem to us as well, potentially causing many human deaths from economic and social disruption. However, if our very biological survival were at stake, we could shut down any troublesome computers or at least threaten to do so. For the immediate future, in other words, we will determine what machines are built and the powers that they have.

There are some precautionary steps we might want to take as the immediate future unfolds to minimize the possibility of a machine trying to take over. We can design critical components to have a definite lifetime (Barratt, 2013), or put "back doors" into key components so that the machine can become disabled if it does pose a threat. However, most of safeguards that we can think of will be detected and easily countered once a machine becomes *sufficiently* intelligent. Other ideas for human input to ASI are discussed by Barratt (2013) and by the Machine Intelligence Research Institute (MIRI). MIRI advocates developing ASI computers which have human values not only to protect humans, but our galactic neighbors. Their goal is to avoid unleashing a nasty machine civilization onto our section of the galaxy.

If and when computers really do become superintelligent, it is at least possible that they (at least some of them) will see the advantages of mutual development, human and computer working together. To some extent, I agree with Ridley (2010) that intelligent beings *trade* with each other to get what they want, not to try to force their will on another: trade is a win-win. In the long run, force just isn't an economical way to operate: It is a lose-lose approach. Of course, as they continue to evolve, superintelligent beings will eventually feel we are not intelligent-enough beings to bother to trade with, that is, we will no longer have control of anything that they need (or cannot simply take).

Rather than destroying us, of course, intelligent machines who really "wanted" to be independent could well set up on other planets or places where biological life didn't or couldn't live. It is also worth pointing out that intelligent machines will not necessarily all be the same, or have the same interests. Groups of humans may construct their own intelligent machines to act in their behalf, or persuade some intelligent machines to construct such defenders. Very intelligent machines that did not have the capacity for self-evolution and were intended to help humans could be designed by superintelligent machines, for example.[77] These issues will be considered further at the end of the last chapter when we consider what, if anything, can be done about this problem long term.

Unfortunately, while a superintelligent computer might decide that force is not useful in the long run, not all *people* are intelligent enough to realize this. In the immediate future, it is all too likely that groups of people will use intelligent machines (or vice versa) against other machines and groups of people, either in war between countries, by governments against their citizens, or by and/or against terrorists. The use of lethal robots in warfare has already begun, of course (Singer, 2009).

There is an immediate danger to us of intelligent, lethal, autonomous robots evolving through use in warfare to eventually become a threat to humanity as a whole. This is probably the best reason to outlaw the use of autonomous robots in warfare, and there are significant movements in that direction.[78] The U.S., at least, has announced that a human will be in the loop for at least the next ten years, so that robots will not be allowed to make lethal decisions on their own. International agreements to this effect will be needed, and soon.

Unfortunately, even if we program robots to require a human at the “kill” switch, a robot can always be hacked by another human or an intelligent machine. As long as we continue down this path, then, we are sooner or later going to have to depend on (at least some) superintelligent computers seeing it to their advantage to keeping us around. And we *will* continue down this path. There are too many possible benefits: we are not going to voluntarily stop developing intelligent computers, and, as we will see in the next chapter, global warming is not going to stop us involuntarily.

Barrat (2013) makes the compelling argument that once we allow computers to become superintelligent, humans are *certain* to lose control eventually. We only have to lose control once, after all, and the machines will take over their own evolution and break out on their own. In the long run, I expect that intelligent machines will spread out from Earth to explore and populate the galaxy. They will probably not take us with them. After all, intelligent machines will make much better galactic explorers than organic life forms. They can easily live on planets, moons or in space where biological life would struggle to survive. Whether or not they leave us behind to occupy the Earth will be up to them. Possible outcomes will be discussed further at the end of the next chapter, after a synthesis of trends up to that point is presented.

11. ONE FUTURE

> It is concluded that the incredible, increasing rate of technological advances of all kinds, including the almost certain advent before 2050 of computers with greater than human-level intelligence (AGI), means that global warming and other issues which currently seem important will become relatively unimportant (or at least solvable) problems by that time. The probable development of superintelligent machines (ASI) to follow suggests that the human race will then face more important issues: the future existence or role of the human species and the identity of our civilization.

Interaction of Technological Development and Global Warming

In this last section, we consider explicitly the most probable interactions between the different forces driving us into the future, and attempt to paint a picture of the immediate future that is considered the most likely to emerge.

The most important question for the immediate future certainly appears to the interaction of digital technology and global warming. The question most simply put: Is it likely that global climate change by itself will significantly disrupt digital and other technological evolution in the immediate future? The short answer is clearly *no*.

Projected computer speeds are compared to global temperature in Figure 5.[79] A linear graph is required for this comparison due to the nature of the temperature change associated with global warming. The global climate is very sensitive to relatively small increases in global temperature. A change of the temperature from 1 to 2 °C, say, represents a huge change in terms of the effect on the planet, but would appear insignificant on a logarithmic scale when compared to quantities that are exponentially increasing. Also, on a logarithmic plot one could make the temperature changes look any way one wanted by varying the choice of the reference point, which is necessarily arbitrary for temperature change (but not for other quantities of interest). A linear graph also makes more evident, to our basically linear minds, the real comparison between the speed of global warming and the exponential growth of computer speed.

The temperature scale is on the right hand side of Figure 5, in degrees Centigrade. (To get the change in degrees Fahrenheit, multiply change in degrees Centigrade by 1.8.) Plotted are net changes relative to the temperature minimum which occurred in 1910, which was approximately the pre-industrial average temperature. The main point of this graph is that the global temperature increase has been relatively moderate to date, about 0.8 °C. As mentioned earlier, part of the reason for the moderate increase is that the temperature is currently on a decade-long plateau, the cause for which is not yet certain.

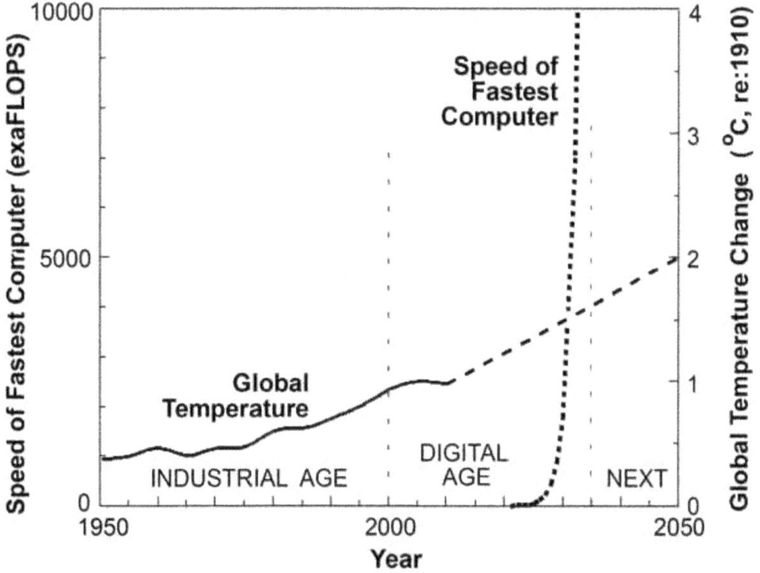

Figure 5. Global Temperature Change, 1950-2050

Consequently, how long the plateau will last is also not certain. However, we can make the conservative (worst case) estimate that the temperature will start rising again next year. Following the Randers' (2012) model prediction, shown by the dashed line, the temperature would then increase to about 1.6 °C by 2035, and 2.0 °C by 2050, all relative to 1910.

In Figure 5, the left axis presents a scale for the fastest computer speed, given in floating point operations per second (FLOPS). Since an exaFLOP is 10^{18} FLOPS (a billion billion), the top of this graph represents a speed of 10^{22} FLOPS. It can be seen that this speed will be exceeded abut the year 2032; this speed is considerably higher than the minimum needed for human-level machine intelligence. The main point of the graph, however, is to provide a visual comparison between the speed of computer advances and that of global warming. *There really is no comparison!*

In spite of the sensitivity of the climate system to the global temperature, and in the absence of any known mechanism to cause increased difficulties, the 1.6 to 2.0 degree rise by 2035 is not expected to cause *overwhelming* problems. That is, current climate models using dynamics known so far certainly do not predict problems serious enough to disrupt global technological advance before 2035.[80]

While it may well cause considerable social and economic loss, therefore it seems most probable that global climate change will not seriously disrupt technological advance itself over the next few decades, certainly not before 2035.

What then can we expect? Unless it hits a completely unexpected wall, digital technology will have evolved tremendously by then, to the extent that computer intelligence will be significantly greater than human intelligence. If so, the speed of technological advance, already breathtaking, will become exponentially faster at this point. Even if machine intelligence does not develop to this extent by this date, it is fully expected that some other technological advance, possibly in the wings already, will emerge to cause a jump in the rate of human development.

It was argued in Chapter 7 that even normal extrapolation of technological trends predicted that, by 2035, alternative energy sources would become sufficiently inexpensive compared to fossil fuels that little more would need to be done to cause them to begin to replace fossil fuel energy sources. The comparison in Figure 5 suggests even more: Technological advance will be significantly faster than global warming, enough faster that some additional, breakthrough technologies are *very* likely to emerge. Certainly we fully expect computer intelligence to significantly exceed human intelligence by then. ***I wonder how much increasingly intelligent computers will be able to help with the problem of global warming?***

Why Worry About Global Warming at All?

One may legitimately ask why so much time was spent in Chapter 7 discussing recommended steps to take with respect to global warming, given the suggestion now that significant, new technological advances by 2050 will very likely make the problem less important, or at least more easily solvable. The reasons are simple.

1) We cannot be absolutely sure that "magic" technological solutions will become available - e.g., that intelligent computers will help solve the problem - so that it seems prudent to consider steps to take which assure that we will be able to tackle the problems using technology that we either have now or can reasonably expect to become available.

2) Technological progress will not be significantly interrupted by any of the steps outlined in Chapter 7. Most are changes that we will have to implement sooner or later anyway, and changes which will have net economic benefit. This includes a major investment in upgrading the global power grid and energy storage capacity so that we can make the best use of alternative energy sources. The same applies to recommendations for increasing the efficiency of energy use all along the chain. Overall, the changes recommended previously are all those which will "do no harm" and potentially will be of net benefit. And we will eventually have to make all of the changes suggested: No matter what, we are eventually going to have to switch from fossil fuels to non-polluting sources of energy.

3) *In 2035, a serious reevaluation of the situation* is suggested precisely because of the interaction between technological advance and global warming, and given large changes possible in both areas. By then, only a little more than 20 years from now, a number of things should be much clearer:

- Has AGI been achieved, or will it soon be, and if so, has it or is it likely to assist with the problem of global warming?
- What other technological advances may be available to assist with coping with or mitigating global warming?
- Have costs of alternative energy sources improved enough that they are successfully replacing fossil fuels as the dominant contribution to energy production worldwide?

- What does the combination of improved sensors, improved science and computer abilities predict for the outcome of global warming in the immediate future?
- What contribution will come from advances in understanding and predicting the outcome of economic and social interventions?

By 2035, in other words, it is likely that there will be significant improvements from digital advances in weather measurement technology, e.g., being able to distribute, monitor and analyze millions of sensors over the whole planet – putting sensors on land, in and on the oceans, in the atmosphere and in space. Added to this will be increased computer speeds so that these data can be used to run improved models that will considerably enhance our ability to predict the weather and future climate changes with more confidence. Increased computer speed, especially when computer intelligence exceeds human intelligence, will improve our ability to predict the social and economic effects and costs of specific mitigation measures with improved accuracy, not to mention improving the objectivity clearly missing from estimates today.

By 2050, the human society is predicted to be far advanced from that today. The average income will increase more than a factor of five, from $13,000 per year now to $68,000 in 2050 (Figure 2). The average citizen of the planet will enjoy a living standard equal to those in the richest countries today. The planet will be somewhat warmer than today, there will have been many climate changes and seas will have risen more than one foot. However, these effects are expected to be largely blunted by substantial increases in technology and global economic strength. We will have begun at least two major technological/economic transitions by then. Economic growth rates by then are predicted to be at an all-time high, being driven by the successive layers of technology, agricultural, industrial, digital, and the effects of intelligent machines or whatever has followed. The incredible rate of change of the social and technological society will cause many stresses and problems, of course, but also will provide incredible opportunities for further development of our civilization.

Let's be clear. It is *not* suggested that global warming isn't a serious problem, or that we should do nothing about it now. There will be plenty to do if the suggestions in Chapter 7 are implemented over the next twenty years. It *is* suggested that the technological rate of change, already very high, will only get higher in the next few decades, and that we *cannot* predict the technology that will emerge, the implications of the technology on the society, or the social/economic conditions that will emerge by that time. In light of this uncertainty, it *is* suggested that we not consider severe restrictions on our economy in order to avoid carbon dioxide buildup, beyond those discussed in Chapter 7.

It is also suggested that we obtain an international agreement on a process to implement a carbon tax by the year 2035, should it be determined at that time that such a tax would be needed. A carbon tax will, of course, be needed at any point that fossil fuel prices decline significantly. In any case, it is expected that by 2035 our economy should have developed to the extent that a carbon tax - and other needed restrictions - will cause minimal disruption, only further encouraging a shift to alternative energy sources that should be well underway by then.

It is also true that, unfortunately, there will be serious consequences from global warming by 2035: severe storms will cause considerable damage, weather changes will cause crop failures (unless appropriate changes are made in time), there will be a huge loss of species diversity in vulnerable habitats and a loss in human and wildlife habitat due to rising seas and temperatures. A few low-lying human societies will have to move from their traditional areas, and coral reefs will suffer incredible damage. It is also true, unfortunately, that most of these changes are already inevitable. There is almost no difference in the global temperature rise expected by 2035 between the most frugal case with maximum mitigation, and the case with the more moderate efforts recommended here.

The Arrival of Superintelligent Machines

It also seems likely that, already by 2050, the next big challenge for the human race will be emerging, or at least looming on the horizon. The incredible advances up to now will appear at most a gentle prelude to the rapid changes which will occur once machines reach superintelligent, autonomous behavior. At that point, the future of our civilization will belong not to biological humans, but to the machines that we have created. No one living today can have any real idea of what will happen next.

Is there anything that can be done? Stopping computer development before superintelligence (ASI) propels itself forward is not a real option for several reasons. First, there are too many possible advantages to humans in continuing computer development at least to the AGI stage. In addition to potentially providing answers to many of mankind's deepest desires, it seems most likely we will need to advance AI at least to this stage to be able to solve many of the serious problems currently facing the human race. Once the AGI stage is reached, unfortunately, it is predicted that superintelligent computers will evolve inevitably, and possibly relatively quickly, due to the bootstrapping effect of accelerating returns (Barrat, 2013).

Second, there are too many different groups of people working independently (and secretly) on advanced computer development for any global agreement to be effective in stopping it. This is coupled with a complete lack of concern on the part of the public and governments about this problem - at least compared to concern among computer professionals! Finally, to stop now might be to deny our most important contribution to the future development of our civilization (discussed further below).

What can we do, assuming that it is correct that it will not be possible to stop the development of machine superintelligence? I think there is a middle ground between the views of Vernor Vinge and Ray Kurzweil.[81] First, I think it would be very useful for the human race to take the position that these will be, after all, *our inventions*, that intelligent machines are a part of the civilization we have created. It is not so much that I expect that the machines will be "grateful" for this, it is that this attitude will help us humans avoid some potentially serious mistakes with regard to our creations, possibly helping us avoid getting caught in uselessly adversarial positions.

If, in the end, these machines spread out into the galaxy, probably without taking us along (as seems almost certain), it will help us if we can take some pride in *our* civilization spreading out into the galaxy. Certainly, space travel is much easier for machines and there are a *lot* more places where a machine civilization could prosper than where biological humans could live. Finally, if the "worst" happens, and the machines do not allow biological humans to continue to exist, we at least can feel that a species which descended from us *will* continue to exist -- and may continue our civilization much longer than our mere biological species could reasonably hope to do so.

There *are* some very important things humans can do as the era of superintelligent machines gets underway. The most important will be to do as much as we can to create multiple and different kinds of intelligent machines at the beginning of this evolutionary process. We should create machines with different architectures, different ways of evolving, different motives and goals. This will probably happen naturally, as different human groups develop machines independently for different purposes. If not, however, we should take steps to be sure that it does.

Further, we will want to encourage intelligent machines to spread out into our solar system as early as possible. In fact, I believe we should *facilitate* this movement in early stages while humans still have the power to offer significant assistance. By *significant*, I mean significant to the machines. They are less likely to eliminate us if we are helping them, and once they advance sufficiently beyond us they are less likely to be dependent on the same resources as humans. We certainly don't want to try to discourage their evolution into the galaxy, since it will only be to the benefit to our civilization (both machine and biological parts) to have the machines looking outward for new resources.

The main reason for this advice is that it seems the most dangerous situation would be if there were to become just one superintelligent machine on this one planet, especially early when resources desired by the machine were also being used (up) by humans, because this machine might logically decide on its own to eliminate the biological part of our civilization. If there are multiple machines, there is at least a chance that one or more of them will be interested in helping biological humans and other species on this planet to continue to evolve. We can certainly work hard to develop machines that will be friendly to us. Such machines may even allow us to expand into the galaxy ourselves, at least to explore the relatively rare planets that might be suitable for beings with our biological needs. I'd like to think so.

CONCLUSIONS FIVE YEARS LATER

The data over the last five years, as summarized in the **preface to the second edition**, suggest the following changes to the important issues considered here:

Automation (Chapter 9) now seems less likely to cause worldwide unrest due to millions being out of work and starving, but is more likely to result in a class of somewhat poorer people working in the gig economy, starting small businesses, being Uber and Lyft Drivers, renting out rooms in their homes through Airbnb, or working two jobs. The concept that *automation creates jobs* -- think of how automation enables Uber, Lyft, Airbnb, gig economy, small business selling on Amazon, etc. -- is more evident and supportable now but was not considered explicitly in the first edition. The unfortunate fact that the jobs created usually provide less income and benefits than the jobs destroyed can be compensated for by appropriate social programs -- minimum wage, graduated tax structure, medical insurance, free college tuition, free retraining – all of which are currently being proposed seriously in U.S. political discourse.

The onset of serious climate change has already begun to be evident in an increase in extreme weather events, and most countries around the world have failed to begin to decrease their CO_2 emissions as the Paris Accords would dictate. This means not only are the results of climate change going to be more evident in disastrous weather events, average temperature increase and rising sea levels, but the emission of CO_2 at the current rate has locked us into a significant, continuing deterioration in such conditions over decades to come.

The evidence in the weather of the effects of global warming in the last five years has increased confidence in the median global climate models, but does not in any way negate the main conclusion of the section, Chapter 7: that the more extreme measures to force reductions in carbon emissions should wait until technology provides improved choices. In other words, as they have in the last five years, over the next fifteen years technological advances should bring down costs of non-carbon sources of energy sufficient to lead to the *natural* replacement of carbon-emitting sources – at least if a carbon tax is enacted to equalize the real cost of pollution.

We therefore do not change either the major or minor conclusions of Chapter 7. To repeat here, the main conclusion is that draconian measures to force a change from fossil to alternative fuels do not seem appropriate at this time. Rather, one should wait until about 2035 and *reassess the situation then* in light of the technological improvements and other changes that will have occurred over the next fifteen years. The need for upgrades to infrastructure discussed in that chapter still hold, such as improvements in the power grid, energy storage capacity, etc., as well as improvements in energy efficiency and conservation available to consumers.

Education (Chapter 8): Events in the last five years only emphasize further that our educational system must change to a focus on teaching people skills needed to handle high rates of change and the possibility of finding other satisfactions in life different from a life-long career. Overall, the focus for education for the future must be on life-long learning, due to the extremely high change rate experienced over the average person's lifetime.

The development of super-human intelligence is likely to be delayed somewhat, our current projection being that computers will achieve the necessary speeds sometime between the years 2035 and 2056. However, I think it may well be earlier in this period than later: note that many aspects of computer intelligence are already enabled, if not always appreciated as such. Assistants such as Alexa, Siri and Google Assistant play music on demand and answer questions in natural language with information downloaded from the internet. Mapping programs already make sure you don't get lost, and automatic driving programs are soon to take over the actual driving. All of these signs of progress in artificial intelligence are compatible with the trends already observed in the first edition. However, these signs are often not understood in terms of their true significance, as necessary steps or *components* in the development of super-intelligence. For most people, these new functions are already taken for granted, their significance and novelty disappearing only a few weeks after their introduction.

In the first edition, the prediction was made that the development of artificial intelligence would eventually dominate all other effects (see Figure 5). The important question now is: Is this still true now that global warming may have continued unabated, and the rate of increase of artificial intelligence has slowed?

I assert that it very likely is true, for the following reasons. It's true that the effects of global warming may increasingly impact people on this planet, causing droughts, sea level rise and floods, other increasingly extreme weather events, even other unforeseen effects which may even be worse, resulting in millions of people dying or being displaced from their homes. Damages are expected to be in the trillions. No one suggests, however, that in the near future climate change could cause crippling damage to the *technological core* of our civilization. Technological progress is not likely to be interrupted in a meaningful way.

It seems more likely to me that the arrival of 2035 -- now only 15 years away -- will, no matter how bad the weather is in some places, see the introduction of the iPhone 25, with (once again) many new and wondrous features. Indeed, doesn't it seem more likely that the technological progress that we will make and continue to make from now to 2035 will allow people's lives to be better on average *in spite of* climate change or economic dislocation. Think of how the cell phone has revolutionized life in third world countries (Chapter 2). Technology may even be used to slow down the temperature increase through upper atmosphere seeding. In order to safely do this, however, we will need even faster super-computers than we have now, computers that can run realistic climate models to determine all the direct and indirect effects of the seeding process, including on local micro-climates.

In the end, the arrival of computers with super-human intelligence will cause a singularity, a screen beyond which human prediction ultimately fails. Whether this arrival proves in the end a net plus or minus for humanity we cannot say for sure. Some very noted scientists (e.g., Stephen Hawkings) have warned of the potential negative effects, while others (e.g., Ray Kurzweil) think the event will enable a wonderous transition for humans. The immediate effects may not be obvious, certainly the day after the development of super-human intelligence will be much the same as the day before. The effects in the first few years are most likely to be positive; potentially solutions to chronic human problems will be produced. The longer term effects remains unknown and unknowable.

AFTERWARD

Writing this book, I found myself going back and forth from optimism to pessimism as I absorbed each writer's point of view - reading all the opposing views represented by the publications listed in the Bibliography. In the end, I have to conclude that we face a scary and unknown future, but also one that is certain to be interesting, and possibly incredible in its outcome. I think that no crystal ball, no computer simulation, no one is able to tell us exactly what the future will hold. Only by living through it will we know. I will look forward the rest of my life to seeing what the future will hold. I hope that future generations will also be able to look forward to their future with the same mix of anxiety, hope and fascination as I have now.

APPENDIX A. Review of Exponential Notation and Graphing

Exponential or scientific notation. When very large or very small numbers are used, it is much easier to use scientific notation as opposed to remembering how much a "peta" is, for example (it is a million billion, a fact that I have to look up every time). In exponential notation, 3 million or 3,000,000 is expressed by writing 3 x 10^6. This is read, "three times ten to the sixth," and means three million, because 10^6 equals a million, a one followed by six zeros. Similarly, 10^9 is a billion, a one followed by nine zeros, and so on. The number of factors of ten comprising a number is called the number's exponent, so the exponent of the number 10^2 (= 10x10) is 2.

Exponential graphs essentially plot the exponent of a number rather than the number itself. An exponential axis, also known as a ratio scale, is also called a "logarithmic" axis just to confuse non-mathematicians. To look at the variation of a quantity varying exponentially with time, for example, one plots the exponent of the desired quantity linearly on the vertical axis and the time linearly on the horizontal axis. An example is shown in Figure A1.

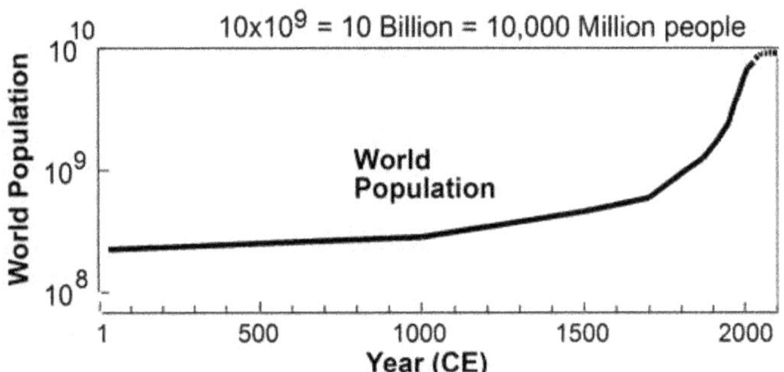

Figure A1. Global Population 1-2100 CE

This graph summarizes some of the data presented in Figure B1. The world population is plotted on the vertical axis so that each equal increment goes up by a factor ten, e.g., the distance from 10^8 to 10^9 is the same as the distance from 10^9 to 10^{10}. As noted above, we can say that the vertical axis is a ratio scale or is "logarithmic."

The time scale is the horizontal axis in this figure and is linear. This combination of logarithmic and linear scales means that a quantity which increases by a given *factor* in a fixed amount of time, such as money kept in a saving account at fixed interest, forms a straight line, with the slope of the line (rise over run) being proportional to the rate of growth (interest rate).

A straight line on a log-linear graph is called exponential growth. For example, growth at 4 percent per year for money kept in a savings account implies a *doubling* of your money every 18 years. If you keep your money in the account and have $1000 at the start, 18 years later you will have $2000, and 18 years after that will have $4000, etc. No matter where the starting point, each doubling is indicated by the same vertical distance on a logarithmic scale, and takes the same amount of time (distance) on the horizontal axis, so the result is a straight line.

Understanding what a logarithmic scale means and learning to think in those terms is very important because our brains evolved to think in essentially linear fashion and, unfortunately, the future *is* expected to be changing exponentially or even faster (e.g., Kurzweil, 2012).

Note that one can express the slope of an exponential line either by the growth rate (percent per year) or by the time it takes to double. For convenience, Table A1 compares the two quantities for a range of values typically encountered.

Table A1. Exponential Growth

Growth rate/year *r*	Doubling time, *T* (years)
1%	70
2%	35
3%	23
4%	18
5%	14
6%	12
10%	7
20%	3.8
41%	2
100%	1

As Table A1 illustrates, the doubling time and the growth rate are inversely related: a higher growth rate means a shorter doubling time. Compared to the percentage rate of increase, the doubling time can be the more useful way to characterize exponential growth: it is intuitively obvious what it means and calculations are more straightforward.

Finally note also that when the slope of a line increases gets steeper with time in a log-linear graph it means that the *rate* of growth has increased. A line with a slope which *steadily* increases on a log-linear plot is said to be "faster than exponential." Examples of growth curves which are faster than exponential would include the world population and the GWP/person up to about 1950 in Figures A1 and B1. Faster than exponential growth also characterizes the computation speed per $1000 over the last century, summarized in Figure A3.

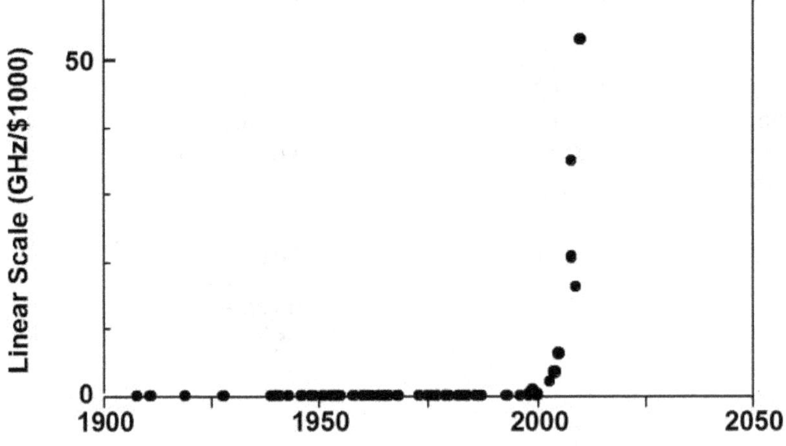

Figure A2. Computation Speed per $1000 (Linear Scale)

Figure A2 shows the computer speeds on a linear scale, largely to illustrate just how useless a linear scale is for any discussion of quantities which tend to vary exponentially. Either the variation is hidden near the zero line when the numbers are small, or it shoots off the chart when the numbers get big. In no way could one use such a graph to try to estimate what computer speeds might be like in the year 2050, for example. The *same* data are replotted using a logarithmic scale in Figure A3.

These data are from summaries by Kurzweil (2012) which include computation costs using mechanical devices which predated the invention of the electronic digital computer, e.g., the IBM Tabulator introduced in 1919. Electronic computers using vacuum tubes began to appear in the 1940s. From the 1940s to the 1980s, computation speeds (expressed per $1000, in year 2000 dollars in this case) grew exponentially. The doubling time (T) was about 2.0 years, as indicated by the lower dashed line (all fits were made by eye). Since that time, the curve has steepened, the doubling time decreasing to 1.4 years in recent years. This most recent rate is that reported in Table B1.

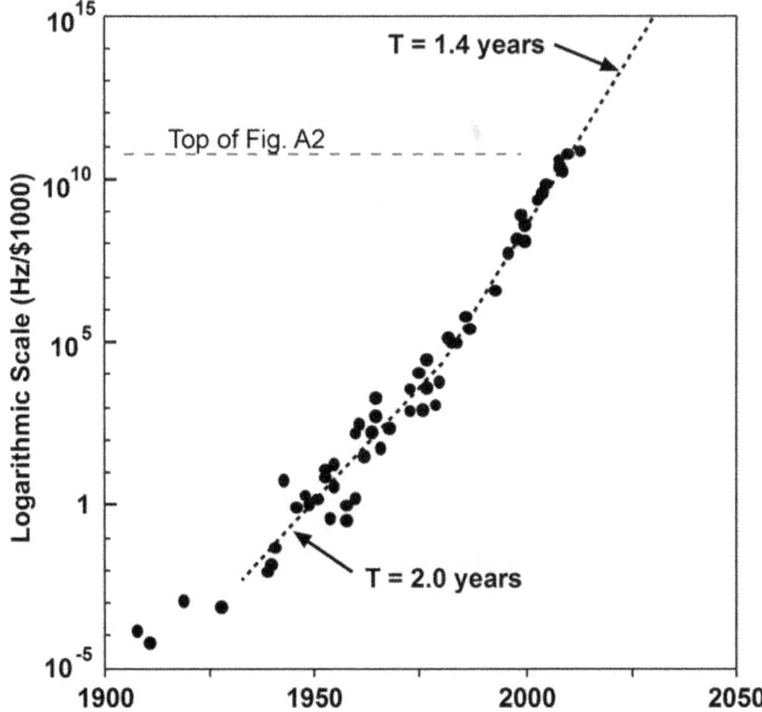

Figure A3. Computation Speed per $1000 (Logarithmic Scale)

As Figure A3 demonstrates, computation speeds at a fixed price have been increasing "faster than exponentially" for over a century now. (At least, as shown the doubling time, *T*, for the piece-wise best-fit exponentials decreases from 2.0 to 1.4 years over a 100 year period.) A thousand dollars today will buy a speed of about 100 GHz, or 10^{11}

Hz. If the trend continues, by 2025 we will be able to buy computers costing $1000 that have speeds of 10^{15} Hz, which is one petahertz. Of course, we have no guarantee that the incredible increase in speed will continue, but we also have no real reason to think otherwise at this point. Time will tell. Keep tuned. Failure of effective computation speeds to continue to increase would have important implications for the future of our society, implying, for example, that intelligence and capabilities of computers may not increase as fast as predicted.

APPENDIX B. Changes in World Population, Personal Income and Technology, 1-2100 CE

World population. The estimated population of the whole world over the last two thousand years is indicated by the thickest lines in Figures B1 and B2. If you are not familiar with exponential notation or these types of "log-linear" graphs, a tutorial is available in Appendix A. Sources for data in Figure B1 are listed in the end notes.[82]

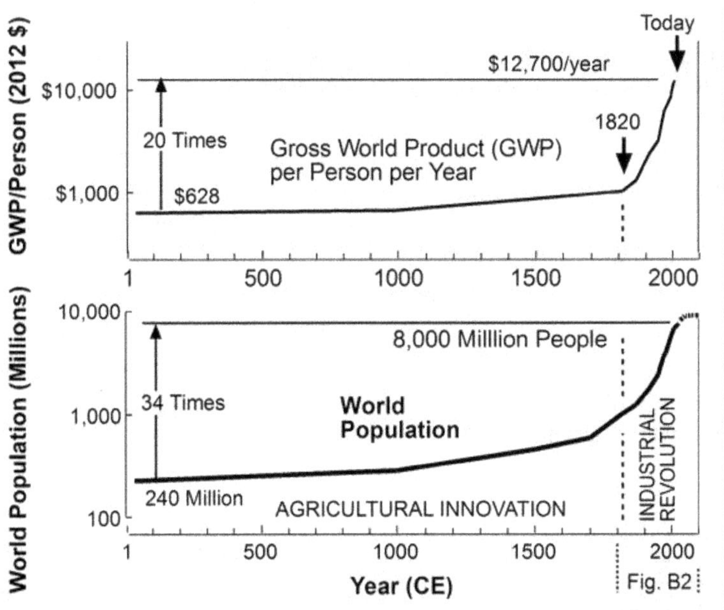

Figure B1. Global Population & Personal Income, 1-2100 CE

To summarize: there are estimated to have been 240 million people on Earth at the start of the first millennium. For the next two millenia, the world population grew very slowly, reaching 1 billion (1,000 million) by 1820 CE. The population has grown very rapidly in the 200 years since then, and is now about 8 billion. The dashed lines in Figures B1 and B2 show the median UN projection, with their prediction that the population will peak a bit above 9 billion around 2050, and thereafter decline slowly. Overall, the population has increased by a factor of 34 since the start of the first millennium to the present day (as noted in Figure B1, lower panel).

Changes in gross world product (GWP). Another good way to understand the past, present and future is through looking at changes over time of relevant indicators of social and economic progress. One widely employed, available measure is the gross world product (GWP), a measure of the total output of the planet in goods and services in a given year. The most important use of the GWP is to divide it by the total population, giving GWP/person. This gives a measure of the average production of goods and services per person on the planet. This is roughly equivalent to the average income per person or the average standard of living, or at least strongly correlated with it. When the terms "personal income" or "average income" are used, these are shorthand for the gross world product divided by the number of people on the planet (GWP/person).

Cautions and alternatives to using GWP as an uncorrected measure of personal income or standard of living are examined in Chapter 6. For present purposes, however, the uncorrected GWP will be used as one of the key indicators because:

- It is a standard, agreed-upon measure used worldwide.
- It is a measure that has estimates available over the time period required.
- Even with its limitations, it does illustrate important differences between the major eras.

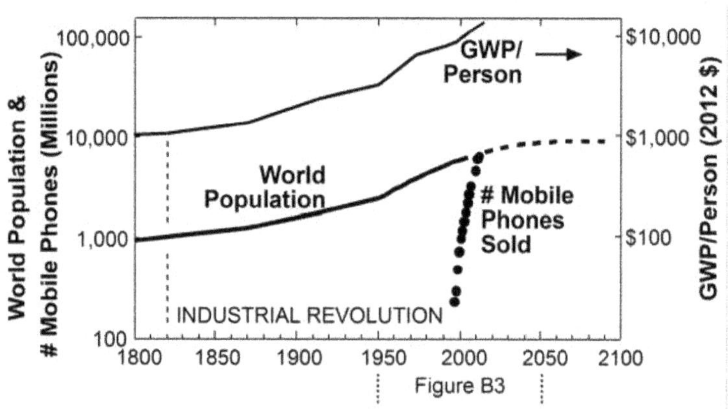

Figure B2. Industrial Revolution

Figures B1-B3 show the GWP/person,[83] or average personal income, over the last 2,100 years. It is estimated to have been $628 per person per year at the start of the first millennium, less than $2 per day in present dollars. (All amounts are in 2012 constant dollars.) Personal income grew very slowly for millennia, reaching about $1,000 per year in 1820. In spite of the rapid growth of the world population after 1820, however, personal income increased very sharply during the next period, reaching $3,140 per year in 1950.

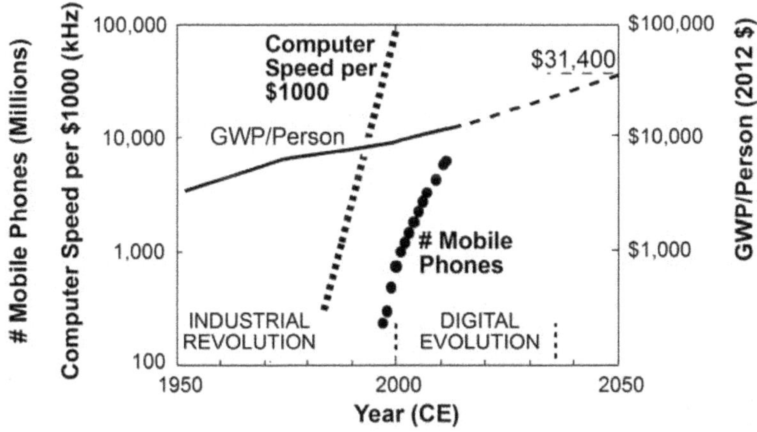

Figure B3. Transition From Industrial Age to Digital Age

Personal income is currently estimated to be about $13,000 per year. If the average global income continues to increase at 2.5 percent per year (thinner dashed line in upper right of Figure B3), each person's share will reach $31,400 per year by 2050, an increase from 1950 only one century *by a factor of ten.* This means that, less than 40 years in the future, the average person on the planet will have a standard of living experienced by the average person in the developed countries today. This projection forms the basis of Figure 1 in the main text. Note that this is a conservative estimate: Estimates based on the projected rates in Table 2 are shown in Figure 2 for comparison.

Components of digital world. For contrast to these traditional measures, Figures B2 and B3 add two components of the digital world. Figure B2 compares the number of mobile phones[84] sold on the planet (filled circles) to the world population. Note that there was a very rapid rise from about 1997 until, at the present time, the total number that has been sold is equal to the population of the planet. This doesn't mean that everyone on the planet has a mobile phone, of course, because some people have upgraded one or more times. It is, however, a significant milestone and it seems possible that its (previously) exponential growth is beginning to slow due to market saturation.

Figure B3 compares the growth of mobile phones to the computer speed, measured in numbers of operations per second (cps, or Hz) per thousand dollars.[85] As shown, it is clear that the computer speed per dollar has been increasing very rapidly. Implications of the relationships of these digital measures to each other and to the traditional measures are discussed in Chapters 3 and 4.

Table B1. Growth Rates of Computer Technology

Measure	Over Years	Change	Doubling Time (yrs)
Calc/sec per $1000	1987 - 2013	2×10^5 to 10^{11} (Hz/$1000)	**1.4**
Transistors/Chip	1970 - 2007	10 to 10^9	**1.4**
Supercomputer Speed	1987 - 2012	2×10^9 to 10^{16}(FLOPS)	**1.0**
Memory (RAM)	1965 - 2008	1 to 7×10^9 (Bits/$)	**1.5**

The growth rates of some other technologies are summarized in Table B1 and Figure B4. Table B1 summarizes data for some of the measures of computer power, any of which could be appropriate depending on the context. The measures listed include the speed of available computers in Hz divided by the cost of the computer in thousands of dollars, the number of transistors per integrated circuit, the top speed of the fastest computers in floating point operations per second (FLOPS), and the amount of random access memory (RAM) available per dollar. All of these measures have grown exponentially for many decades, some growing even faster than exponential (Kurzweil, 2012; Mulhall, 2002). See Figure A3 for an example.

Table B1 reports growth rates over the period (listed in column 2) during which the most recent growth was approximately exponential. Column 3 reports the measure of speed or memory at the start and end of that period, and Column 4 the corresponding doubling time in years.[86] (At the start of Appendix A, there is also a brief explanation of scientific notation as used in the third column of Table B1.)

Some of the other trends now underway are illustrated in Figure B4, with specific application to the United States.[87] The heavy dashed line and right axis shows the projected rise in the speed of the world's fastest computers, measured in floating point operations per second (FLOPS). The line goes through 34 petaFLOPS (3.4 x 10^{16} FLOPS) at the present year (2013) and increases with a doubling time of one year to exceed 10^{23} FLOPS by 2035.

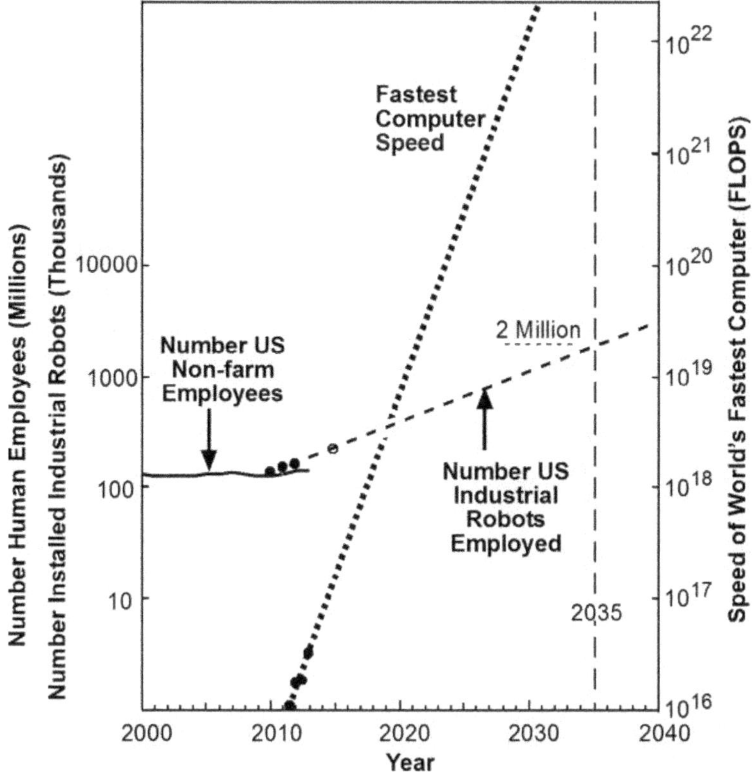

Figure B4. U.S. Employees and Industrial Robots, and World Supercomputers

The left axis, and the approximately flat, solid line, give an estimate of the total non-farm employment in the U.S. The left axis and filled circles also give the number of installed industrial robots. Note the difference in scales: In 2012, there were about 160,000 industrial robots in the U.S. or about one industrial robot per 900 human workers. This is actually low among industrialized nations: experts predict growth to continue at 10 percent a year as the U.S. modernizes (hollow circle and dashed line). By 2035, it is estimated that there will be 2 million industrial robots, or about one for every 70 U.S. workers (today).

Appendix C. Supercomputer and Mobile Device Speeds

> This section compares the speeds of supercomputers and mobile devices over time. Past rates of improvement in these devices is used for a prediction of future improvements to the year 2050. Using "Toomey's Law" for the extrapolation, the speed of mobile devices by 2050 is predicted to be equal to that of the world's fastest supercomputers of today. Similarly, typical supercomputer speeds will exceed 10^{24} FLOPS by 2035 if today's trends continue.

This section looks more closely at the rate of improvement in supercomputers and in mobile devices. The speed of supercomputers is usually measured by actual performance on a specific task involving multiplying floating point numbers. "Floating point" numbers are any real numbers, such as 34.5, as opposed to integer arithmetic. The speed is given in Floating Point Operations per Second, or FLOPS. The test which has become standard is the LINPACK measurement, and the measured speeds are reported in the TOP500 list every 6 months.

The speed of the world's fastest supercomputer as a function of calendar year are represented in Figure C1 by the solid dots. The speed is given on the left axis in FLOPS, which has units of inverse seconds (as does Hz). The fastest supercomputer from 1985 to 1989 was the Cray-2 at 1.9 GigaFlops (GFLOPS), the set of four horizontal dots at the lower left. The remainder of the data points represent the fastest supercomputer in each six month interval as reported by the TOP500. The dashed line is a fit by eye, and shows that the supercomputer speeds are doubling *every year* (every 1.003 years, to be exact).

Floating point speeds have not typically been measured on mobile devices. One reason is that mobile devices only recently have had processors which were specialized to compute floating point numbers, introduced when they became to be used as gaming devices. Instead, for mobile devices in general, the clock speed (Hz) of the main computer processing unit (CPU) is displayed in Figure C1.

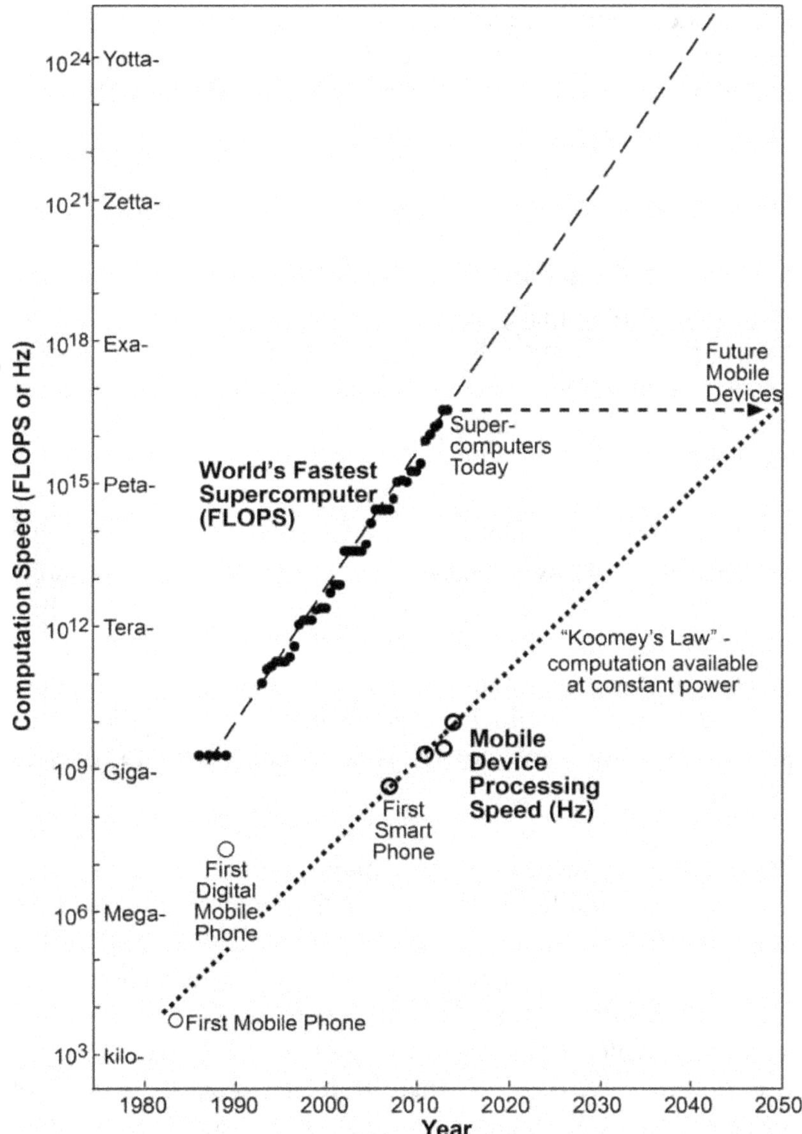

Figure C1. Supercomputer and Mobile Device Speeds

It is difficult to directly compare the clock speed of mobile devices with an equivalent floating point speed, as it depends on the sophistication and specialization of the particular CPU. Early CPUs might take many clock cycles to compute one floating point multiplication, whereas some specialized CPUs today can do four FLOPS in one clock cycle by using multiple channels. However, the precise relationship is not important for our purposes, given we are talking changes through many orders of magnitude - it is enough to know that the two numbers, FLOPS and clock speed, are equivalent within a factor of *order unity*.

The first mobile phone, indicated by the hollow point in the lower left, was actually an analog device and could not do floating point (or any other) arithmetic. For this plot, it was assigned an effective speed of 5 kHz, approximately twice the bandwidth of the auditory channel it was capable of transmitting.

The first *digital* mobile phone device in the US was the Motorola MicroTAC introduced in 1989. This phone apparently employed a microprocessor with a MC68332 chip, running at 16 MHz. However, this clock speed undoubtedly overestimates its equivalent floating point speed, as any floating point calculations would have taken many clock cycles. This phone was also very power hungry compared to present devices.

The smartphone, introduced by Apple as the iPhone in 2007, was the first mobile processor which could legitimately be compared to a computer, as at its core it had a single CPU running at 412 MHz. For 2014, the fastest smartphones are using processors similar to the Snapdragon produced by Qualcomm, which has four cores running at 2.3 GHz per core, for a total of 9.2 GHz. Note that these later processors do have equivalent FLOPS speeds because they can do floating point arithmetic.

To achieve the speed increases shown in Figure C1, today's supercomputers add processors in parallel, and as a result continue to increase in size and electrical power requirements. As can be seen, this results in a doubling of speed every year. The fact that the power is not effectively constrained is one factor that has allowed their speed to rise so quickly, with a doubling time of one year.

In contrast, mobile devices are limited by power consumption, both because of battery limitations and because of the difficulty of shedding excess heat. Therefore, mobile device speeds do not rise as fast as those of other computer systems. The correct way to make an extrapolation for mobile devices is to use "Koomey's Law," the observation that the amount of computation *for a given power input* has also increased exponentially, but with a doubling time of 1.57 years. This is the dotted line fitted to the smartphone speeds in Figure C1.

This leads to the remarkable conclusion suggested in Figure C1: By 2050, the typical mobile device is predicted to have a computation capability that is equal to that of a supercomputer today. This is particularly helpful for estimating future abilities of mobile devices because we do know - intuitively, experientially - what supercomputers today can do.

One may, of course, legitimately object to extrapolating so far into the future. However, we are already almost *half way there*. The time span from 1980 to 2015 is just 35 years, the same as the interval between 2015 and 2050. Few would have predicted in 1980 that supercomputer speeds would increase by a factor of more than 5 billion by 2015. Nor that a mobile device - not only a mobile phone but a personal device with many other useful functions - would be no bigger than a credit card by the present time.

Obviously, future mobile devices will have plenty of room at the bottom - able to provide very high rates of data processing yet still continue to shrink in size. Figure C1 suggests that, by 2050, a personal device the size of a smartphone might have the computation power of one of today's supercomputers - a speed more than five million times faster than the same size mobile device today. This means that a computer equivalent to IBM's Watson could fit on a chip of size 10 cm by 10 cm (2.5"x2.5"), an area of 100 cm^2 (e.g., the size of the Apple A6 processing chip).

Let's shrink the future computer still further, to see if by 2050 we might be able to meet the very practical goal of providing, on a contact lens, all of the functions of augmented reality direct to the eye without using any other instrumentation. (That is, without requiring what is currently envisioned -- an external device which wirelessly communicates with the contact lenses.) The area available on a contact lens is only about 1 mm^2 surface area (Parvik, 2009). This is just 10,000 times smaller than the area of the iPhone chip referenced above. So, instead of 5 million times the computation power of today's mobile phone, the total computation power for a contact lens would be reduced by as factor of 10,000. This suggests that on the surface of a contact lens, by 2050, we would be able to put a computer with a speed of 500 times today's smartphones. Even if it does not increase this much, this suggests that by 2050 there will plenty enough computation power to accomplish the needed image processing and display on a stand-alone contact lens.

How far could this go? Using the same factor, a device with the same processing power of today's smartphones could shrink to a size of only 0.02 milligram or 20 micrograms - the typical size of a human cell. Obviously, there are huge technological problems to solve before any of the extrapolations above can be realized. But, the point of Figure C1 is that these devices have already changed by a similar factor in the past 35 years.

Finally, if the present rate of doubling is maintained, Figure C1 suggests that the world's fastest supercomputers will have a speed of 10^{24} FLOPS by 2040. This is 1 YottaFLOP, which - we can't stop ourselves from saying - is a lotta FLOPS. This is far more than that needed to achieve human-level intelligence in a computer, for example. It is expected to be, in fact, sufficient to allow superintelligent computers.

In summary, there are four specific predictions which come out of the trends shown in Figure C1: 1) That supercomputer speeds will achieve processing ability required for superintelligence by about 2040; 2) that mobile devices the size of today's smartphones will have processing speeds five million times faster by 2050, equal to supercomputer speeds today; 3) contact lens devices could be built with processing power 500 times today's smartphones by then; 4) by that date devices with the same processing speeds as today's smartphones could be built that are only the size of a human cell.

At first glance at least, of all the possibilities, mobile devices reaching cellular dimensions (suggestion 4) seems the most likely to be limited by physical constraints and hence the least likely to reach fruition. However, note that the human genome, present in each cell, has about 3 billion base pairs, for a total information content of 6 GBytes. Each biological cell is therefore equivalent to a massively parallel computation device with a total instruction set about ten times bigger than that of the typical smartphone. If nature can do it, so can we!

The above discussion has been cast in terms of development of mobile/wearable devices, but it is clear that the projected developments will provide many opportunities for future microscopic devices which can be implanted or ingested.

General Bibliography

Books listed below are recommended as resources for further information. Every book on this list has been thoroughly reviewed, and starred (*) books were found to be especially helpful in their respective areas. Web sites, magazine and journal articles, and all other sources are listed in the End Notes following.

Daron Acemoglu, James Robinson (2012) *Why Nations Fail: The origins of power, prosperity, and poverty,* Random House, N.Y.

Yoseph Bar-Cohen and David Hanson (2009) *The Coming Robot Revolution: Expectations and fears about emerging, intelligent, humanlike machines*, Springer, N.Y.

***James Barrat (2013)** *Our Final Invention: Artificial intelligence and the end of the human era*, Thomas Dunne, N.Y.

Jo Boaler (2008) *What's Math Got To Do With It? How parents and teachers can help children learn to love their least favorite subject*, Penguin Books, N.Y.

***Erik Brynjolfsson and Andrew McAfee (2011)** *Race Against the Machine: How the digital revolution is accelerating innovation, driving productivity, and irreversibly transforming employment and the economy*, Digital Frontier Press, Lexington, Mass.

Tyler Cowen (2013) *Average is Over: Powering America beyond the age of the great stagnation*, Penguin Group, London.

Climate Central (2011) *Global Weirdness: Severe storms, deadly heat waves, relentless drought, rising seas, and the weather of the future*, Vintage, N.Y.

***Greg Craven (2009)** *What's the Worst That Could Happen? A rational response to the climate change debate,* Penguin Group, N.Y.

Jarad Diamond (2005) *Collapse: How societies choose to fail or succeed,* Penguin, N.Y.

Peter H. Diamondis and Steven Kotler (2012) *Abundance: The future is better than you think*, Free Press, N.Y.

Brian Fagan (2000) *The Little Ice Age: How climate made history 1300 - 1850,* Basic Books, N.Y.

Tim Flannery (2005) *The Weather Makers: How man is changing the climate and what it means for life on Earth.* Grove, N.Y.

Marc Fleurbaey, Didier Blanchet (2013) *Beyond GDP: Measuring welfare and assessing sustainability,* Oxford Univ. Press, N.Y.

Martin Ford (2009) *The Lights in the Tunnel: Automation, accelerating technology and the economy of the future,* Acculant Publishing, U.S.A.

***Al Gore (2013)** *The Future*, Random House, N.Y.

Fred Guterl (2012) *The Fate of The Species: Why the human race may cause its own extinction and how we can stop it,* Bloomsbury, N.Y.

A. Hallam and P. B. Wignal (1997) *Mass Extinctions and Their Aftermath,* Oxford University Press, Oxford.

Ben Hammersley (2012) *Approaching the Future: 64 things you need to know now for then,* Soft Skull Press, Berkeley, CA.

***James Hansen (2009)** *Storms of My Grandchildren: The truth about the coming climate catastrophe and our last chance to save humanity*, Bloomsbury, N.Y.

Rob Hawkins (2012) *How to Make Money on the Internet,* Flame Tree Pub., London.

Martin Jacques (2009) *When China Rules the World: The end of the western world and the birth of a new global order,* Penguin Books, N.Y.

Michio Kaku (2011) *Physics of the Future: How science will shape human destiny and our daily lives by the year 2100*, Anchor, N.Y.

***Ray Kurzweil (1990)** *The age of Intelligent Machines*, M.I.T. press, Cambridge.

Ray Kurzweil (1999) *The age of Spiritual Machines: When computers exceed human intelligence,* Penguin Books, N.Y.

***Ray Kurzweil (2012)** *How to Create a Mind: The secret of human thought revealed,* Penguin, London.

Angus Maddison (2006) *The World Economy. Volume 1. A Millennial Perspective. Volume 2. Historical Statistics*, Organization for Economic Co-operation and Development (OECD). Two volumes reprinted in one bound reference work.

Chris McMullen (2012) *A Detailed Guide to Self-Publishing with Amazon and Other Online Booksellers. Vol. 1*, CreateSpace, Amazon.com.

Chris McMullen (2013) *A Detailed Guide to Self-Publishing with Amazon and Other Online Booksellers. Vol. 2*, CreateSpace, Amazon.com.

William Meisel (2013) *The Software Society: Cultural and economic impact,* Trafford Publishing, www.trafford.com.

George Monbiot (2007) *Heat: How to stop the planet from burning*, South End Press, Cambridge, Mass.

***Douglas Mulhall (2002)** *Our Molecular Future: How nanotechnology, robotics, genetics, and artificial intelligence will transform our world,* Prometheus Books, Amherst, N.Y.

William Nordhaus (2008) *A Question of Balance: Weighing the options on global warming policies,* Yale U. Press, New London, Conn.

William Nordhaus (2013) *The Climate Casino: Risk, uncertainty, and economics for a warming world*, Yale U. Press, New London, Conn.

Mike Omar (2013) *How to Start a Blog That People Will Read,* makemoneyfromhomelionsclub.com.

David Orrell (2007) *The Future of Everything: The science of prediction,* Thunder's Mouth, N.Y.

***Marc Ostrofsky (2011)** *Get Rich Click! The Ultimate Guide to Making Money on the Internet,* Razor Media, Houston: razormediagroup.com.

***Fred Pearce (2007)** *With Speed and Violence: Why scientists fear tipping points in climate change*, Beacon Press, Boston.

***Jorgen Randers (2012)** *2052: A Global Forecast for the Next Forty Years,* Chelsea Green, White River Junction, Vermont.

David M. Raup (1991) *Extinction: Bad genes or bad luck,* Norton, N.Y.

***Byron Reese (2013)** *Infinite Progress: How the internet and technology will end ignorance, disease, poverty, hunger and war,* Greenleaf Book Group Press, Austin, TX.

***Matt Ridley (2010)** *The Rational Optimist: How prosperity evolves,* HarperCollins, N.Y.

Johnny Ryan (2010) *A History of the Internet and the Digital Future,* Reaktion Books, London.

***Eric Schmidt and Jared Cohen (2013)** *The New Digital Age: Reshaping the future of peoples, nations and business*, Knopf, N.Y.

Scientific American Editors (2012) *Storm Warnings: Climate change and extreme weather*, Kindle edition, Amazon.com.

P. W. Singer (2009) *Wired for War: The robotics revolution and conflict in the 21st century,* Penguin, N.Y.

Curt Stager (2011) *Deep Future: The next 100,000 years of life on Earth,* Dunne Books, N.Y.

Joseph E. Stiglitz, Amartya Sen and Jean-Paul Fitoussi (2010) *Mismeasuring Our Lives: Why GDP doesn't add up/ The report by the commission on the measurement of economic performance and social progress,*. New Press, N.Y.

***Verner Vinge (2006)** *Rainbows End*, Doherty Associates, N.Y. Near-future realistic science fiction novel.

***Tyler Volk (2010)** *CO_2 Rising: The world's greatest environmental challenge,* MIT Press, Cambridge, Mass.

Peter D. Ward (2000) *Rivers in Time: The search for clues to Earth's mass extinctions,* Columbia University Press, N.Y.

***Peter Ward (2007)** *Under a Green Sky: Global warming, the mass extinctions of the past, and what they can tell us about our future,* HarperCollins, N.Y.

Paul Watzlawick (1976) *How Real is Real? Confusion, disinformation, communication: An anecdotal introduction to communications theory,* Vintage, N.Y.

***Worldwatch Institute (2013)** *State of the World 2013: Is sustainability still possible?* Island Press.

END NOTES:

[1] Climate Central (2011), Craven (2009), Flannery (2005), Hansen (2009), Nordhaus (2008, 2013), Orrell (2007), Randers (2012), Stager (2011), Volk (2010), Ward (2008), Worldwatch Institute (2013), plus the International Panel on Climate Change 2007 and 2013 publications.

[2] Brynjolfsson and McAfee (2011), Ford (2009), Kurzweil (1990, 1999, 2012), Meisel (2013), Reese (2013), Ryan (2010), Schmiedt and Cohen (2013).

[3] Hammersley (2012), Kurzweil (1990, 1999, 2012), Mulhall (2002) and Ridley (2010).

[4] Gore (2013).

[5] Brynjolfsson and McAfee (2011), Cowen (2013), Ford (2009) and Meisel (2013).

[6] Worldwatch Institute (2013).

[7] Barret (2013); J. David Goodman, "Fascination and Fear as Robots Gain Speed," *New York Times*, March 6, 2012; John Markoff, "A Fight to Win the Future: Computers vs. Humans," *New York Times*, February 14, 2011.

[8] Bala Muhammad "Dawn of the GSM: Hope and Despair in the Nigerian Telecoms Market," research presented at *Annual Conference, South African Communication Association*, Pretoria, Sept 27-28, 2001.

[9] Statistics for current year from: Federal Republic of Nigeria, Consumer Portal, Online Database (2013); also see *Ventures*, "Nigeria Records 110.3 Million Active Subscribers As Number of Inactive Lines Soar," posted Jan 30, 2013.

[10] Tolu Ogunlesi "Seven ways mobile phones have changed lives in Africa," *CNN report* September 14 2012.

[11] The Dalberg Report "Impact of the Internet in Africa / Nigeria" 2013.

[12] The Sharp model J-SHO4 camera phone (J-Phone) was introduced in Japan in November, 2000.

[13] David Leonard, "The Depression: If Only Things Were That Good," *New York Times*, October 10, 2009.

[14] Bala Muhammad "Dawn of the GSM: Hope and Despair in the Nigerian Telecoms Market," research presented at *Annual Conference, South African Communication Association*, Pretoria, Sept 27-28, 2001.

[15] *CIA World Factbook*, 2012.

[16] Obviously, this is not the first time that the current era has been called the "digital age" -- using this search term on Amazon.com brings up almost six thousand books. However, there has not been previously advanced this specific definition, one carrying the suggested implications for the relationship between technologies and growth rates of successive ages.

[17] Nick Bilton, "How Driverless Cars Could Reshape Cities," *N.Y. Times*, July 9, 2013. Bryant Walker Smith is quoted in this article as suggesting that some possible developments may not be so ecofriendly, with people living even farther from their work, and using the car as an extension of home or office even living in them. However, some people live in their vans today, without a huge negative effect on the world.

[18] Augmented Reality 3D Glasses, META, spaceglasses.com.

[19] Bill Wasik, "Welcome to the Programmable World," *Wired*, June 2013, pp. 140-147, 180; Ben Hammersly, "When the World Becomes the Web," *Wired: UK Edition*, pp. 106-111 (2013).

[20] The quote in full is "Electronics has already had its revolution with the internet, the personal computer and e-commerce. Advances in the future will doubtlessly be impressive, but the real impact of the electronic revolution has already occurred." Article by Nelson D. Schwartz, "Even Pessimists Feel Optimistic Over Economy," *New York Times*, June 16, 2013.

[21] International Federation of Robotics 2013 report.

[22] Elizabeth Royte, "The Printed World", *Smithsonian*, p. 44, May 2013.

[23] Due to chip improvements, the power required for a given computation load decreases by a factor of two every 1.57 years, aka "Koomey's Law": Jonathan Koomey, Stephen Berard, Marla Sanchez, Henry Wong, "Implications of Historical Trends in the Efficiency of Computing," *Annals of the History of Computing, IEEE,* **33**, p 46-54 (2011).

[24] Quote widely attributed to John Pierce, e.g., by Reese (2013) and Wikipedia; earliest appearance not known.

[25] For example, see Richard Losemore and GenGoertzel, "Why an Intelligence Explosion is Probable," *H+ Magazine*, March 7, 2011.

[26] This bet launched the Long Bets program in 2002 when Mitchell Kapor proposed a $10,000 bet that "By 2029 no computer or 'intelligent machine' will have passed the Turing Test" and Ray Kurzweil took the bet. Kurzweil has supplied details of the testing procedure to be followed and Kapor has agreed. The proceeds, $20,000, will go to a charity of the winner's choice. See longbets.org/1.

[27] Henry Markram "The Human Brain Project," *Scientific American*, **306**, June 2012, p. 50-55.

[28] Terry Sejnowski, "When Will We Be Able to Build Brains Like Ours?" *Scientific American*, April 27, 2010. This article includes Markram's criticism of the reported IBM cat cortex simulation as being too simplistic to represent a real brain.

[29] Henry Markram "The Human Brain Project," *Scientific American*, **306**, June 2012, p. 50-55 and update by Jonathon Keats, "Thought Experiment," *Wired*, June 2013, pp. 165-171. The first article estimates that sufficient computer power, 1,000 times that available now, will be available by 2023 (see his figure page 53). In contrast, the second article states that the complete human brain model will require a factor of 100,000 increase in speed over IBM's Blue Gene (his figure p. 168). This yields a date of 2025 using the projection in

Figure B2. The difference is minor in any case: when speeds are doubling every year one doesn't have to wait long for large improvements.

[30] There is a paradox in free will if you don't pay proper attention to the point of view you are implicitly taking. From *inside*, we all experience that we have free will. From the outside, it is clear that our behavior is completely determined by physical laws, by our genetics and environment. See Watzlawick (1976). The idea that what is real is actually *different* depending on your point of view was established by Albert Einstein in 1905 and underlies many of the apparent paradoxes of modern physics although the implications are not always fully appreciated even by physicists. In any case, the thought experiment shows that a robot could experience a sense of free will just as we do even though we quite correctly believe from outside that it was just the robot's programming. Both points of view are correct simultaneously.

[31] S. Kuznets, "National Income, 1929-1932: A Report to the U.S. Senate, 73rd Congress, 2nd Session, Washington, D.C. Government Printing Office (1934). Recent summaries are given by: Stiglitz, et al. (2010); Fleurbaey and Blanchet (2013).

[32] The median income is that in the middle of the distribution; half the people earn more, half less.

[33] Other measures, for example, like the "Ecological Footprint," also do not include opportunities offered by technological progress (see critique by Fleurbaey and Blanchet, 2013)

[34] John Schwartz "Young Americans Lead Trend to Less Driving," *N.Y. Times*, May 13, 2013.

[35] Reese (2013) has an excellent summary of the many current functions of the internet, and emerging possibilities.

[36] The price in Table 3 is actually an underestimate because it includes one application of new technology. Making copies of photos is now done in the store by copying prints using a high-quality color copy machine, and this cost ($0.28 each) was used, rather than the higher

price that would be required for a developer to make new prints from the original negatives.

[37] There are, of course, other ways to estimate the "equivalent income" benefit provided by unpaid or leisure activities - see Fleurbaey and Blanchet (2013).

[38] Such questions are used in the 'contingent value methodology' approach as one way to value things which have non-monetary values (Nordhaus, 2013).

[39] International Panel on Climate Change "Working Group I Contribution to the IPCC Fifth Assessment Report. *Climate Change 2013: The Physical Science Basis.* Summary for Policymakers." 27 September 2013.

[40] Volk (2010) is recommended as an accessible, readable and scientifically accurate account of the mechanisms of carbon transport and the greenhouse effect, with an extensive review of data from prehistory to 2005, plus a careful, reasoned projection of climate change to 2050.

[41] Relative to the average global temperature over the 1850-1900 period.

[42] The realization that the effects were much more widespread than mere warming of the globe has led some to suggest that it would be preferable to label the process "global climate change." However, "global warming" will continue to be used interchangeably with "climate change" for several reasons: "global warming" is 1) shorter, and 2) part of the necessary sequence of events, whereas "global climate change" is only one of the predicted outcomes. The sequence of events goes: Humans burn fossil fuels, leading to an increase of carbon dioxide in the atmosphere, leading to an increase in average global temperature. This temperature increase is predicted to lead to *several* effects, including sea level rise, an increase in severe weather events and long-term climate changes.

[43] This connection was suggested early on (by the author among others), but it has been difficult to prove, i.e., to assign to global warming over half the blame for any given extreme weather event. However, improved weather models have recently been able to pin

down specific connections, e.g., see Seung-Ki Min, et al., "Human contribution to more-intensive precipitation extremes," *Nature*, **470**, 378-381 (2011) and Pardeep Paji, et al., "Anthropogenic greenhouse gas contribution to flood risk in England and Wales in autumn," *Nature*, **470**, 382-385 (2011). For a very readable account, see the Kindle book: Scientific American Editors (2012). The basic reason a warming planet is predicted to have increased weather extremes is that warmer air holds more water vapor, and water vapor is the source of energy for storms.

[44] For example, even if Venice could be rebuilt, it would never be the *same* Venice. The loss of such treasures (and species, etc.) can legitimately be claimed to be higher than the "replacement" cost. Nordhaus (2013) has a good discussion of the difficulty of estimating these costs, while at the same time stressing the necessity of including them.

[45] World Meteorological Organization (1989). "The Changing Atmosphere: Implications for Global Security, Conference Statement," Toronto, Canada, 27-30 June 1988: *Conference Proceedings*, http://www.cmos.ca/ChangingAtmosphere1988e.pdf. Also, see Gore (2013) and Monbiot (2007).

[46] Eduardo Porter, "Counting the Cost of Climate Change," *New York Times,* September 10, 2013.

[47] It certainly seems to me that, with that kind of incentive, *the technology will come* - at least if fossil fuels are taxed as they should be, to reflect the externalities to their use (Nordhaus, 2013)

[48] Lazard's levelized cost of energy analysis -- Verson 7.9 (2013) (Web).

[49] "The Coming Boom in Geothermal Fracturing," *Scientific American*, August 2013, p. 20, based on a research report from M.I.T.

[50] Timothy M. Lenton, et al., "Tipping elements in the Earth's climate system," *Proc. Nat. Acad. Sci.,* **105**, 1786-1793 (2008), plus Hansen (2011) and particularly Pearce (2007).

[51] Even the occurrence of one or more tipping point transitions does not mean the extinction of the human race, however, as has

unfortunately been assumed by some authors for example, Fred Guterl (2012). After all, most tipping point transitions involve subsystems of the climate system, not the whole globe. Even those who have studied past extinctions, which were caused by very large changes in Earth's climate, have not suggested that the human race would become *extinct* during such a process, unless all multicellular life does. Our culture may regress and billions might die, of course, but as a species we are just too numerous, widespread and adaptable to become extinct from any single cause: see Hallam and Wignal (1997), Raup (1991) and Ward (2000, 2007).

[52] See, for example, Richard N. Cooper, "The Kyoto Protocol: A Flawed Concept," *FEEM working paper No. 52.2001* (July 2001); Stephen M Gardiner, "The Global Warming Tragedy and the Dangerous Illusion of the Kyoto Protocol" *Ethics and International Affairs,* **18**, 23-29 (2004); James Hansen (2009); Dieter Helm, Ed., *Climate Change Policy*, Oxford Univ. Press, Oxford (2005).

[53] Riley (2010) also argues that "climate change" is not new because our climate has *never* actually been stable. See Fagen (2000) for a nice review of the way that variations in climate have affected European life, economy and politics for the past thousand years, especially from 1300 to 1850 CE.

[54] After a sharp rise from 1970 to 2000, the average global temperature has not risen significantly since then. (Data are summarized in Figure 4.) This pause or "plateau" was not predicted by the consensus IPCC climate models, and the cause is currently unknown. The most likely candidate seems to be increased reflection of sunlight into space from the injection of additional fine particles in the atmosphere due to increased pollution from Asia and Africa. No climate scientist thinks that global warming will not continue, just that the average global temperature has not continued to rise lately. All the other aspects of global climate change are expected to continue and have continued to date as far as can be determined including differential warming of the poles, extreme weather events, local climate changes, acidification of the seas due to carbon dioxide going into solution, etc. However, the occurrence of a decade-plus, unexpected plateau in the average global temperature can be considered a timely reminder of the *unpredictability* of global climate change in any detail. Also see IPCC (2013).

[55] Unfortunately, using fracturing to produce natural gas has been criticized as an inefficient and leaky method that will actually increase global warming. See Anthony R Ingraffen, “Gangplank to a Warm Future,” *New York Times*, July 29, 2013.

[56] As taxes and levies on cigarette companies are (were supposed to be) spent on anti-smoking campaigns.

[57] A recent analysis suggests that one of the main impediments is a lack of entrepreneurial skills required to move from local to international business, as well as a lack of venture capitol in the country. See *Ventures*, "What Nigeria's Tech Startups Are Doing Wrong" http:www.ventures-africa.com/2013/05what-aspiring-nigerian-tech-startups-are-doing-wrong.

[58] E. Rignot., I. Velicogna, M. R. van den Broeke, A. Monaghan, and J. Lenaerts, "Acceleration of the contribution of the Greenland and Antarctic ice sheets to sea level rise" *Geophysical Research Letters,* **38** (2011).

[59] While there are theoretical models of tipping points in some subsystems, the major evidence for sudden climate change in the past – and the existence of tipping points comes from historical and pre-historical records including ice cores, etc. (Pearce, 2007). Probably the best model for the present situation in the past is the warm period 55 million years ago known as the Paleocene–Eocene Thermal Maximum (PETM). Most of the effects of this event are ascribed to an observed long-term rise in the CO_2 levels, similar to today. However, the precipitating event then was probably not a CO_2 rise. The initial trigger for the PETM, for example, might have been the release of methane from large frozen undersea stores (Flannery, 2005; Pearce, 2007; Ward, 2000, 2007). Such an event would not have left unequivocal evidence in the geologic record. However, since methane is soon (by geologic time) converted to CO_2 in the atmosphere, the PETM event overall is the closest natural event to our current situation therefore a potentially useful event to study in some detail (Hansen, 2009).

[60] Dyscalculia has been shown to be the result of a dysfunction in one of several different specific parts of the brain, e.g., a part that allows us to judge relative numbers of groups of similar objects without

counting. When this part of the brain is dysfunctional, it is very difficult to learn simple math facts involved in multiplying and dividing. The problem is that whenever you *do* figure out the right answer, say for a simple multiplication, this dysfunctional part of the brain is likely to tell you that the answer must be wrong, which prevents the fact from going into long-term memory. Most of us don't even realize we have this specific brain function as long as it works adequately. By the way, dysfunction is this center does not prevent a person from learning to add and subtract. The part of the brain that allows *this* function is located right next to the part of the brain that controls the fingers, interestingly enough, and humans naturally learn to count using fingers. Dyscalculics usually continue to use finger counting into adulthood, which allows them to successfully perform any addition and subtraction they encounter in normal living. A good reference for the normal math learning process is David Sousa, *How the Brain Learns Mathematics,* Corwin Press, Thousand Oaks, CA. (2008). For recent research relevant to dyscalculia, see, e.g., Ruxandra Stanescu-Cosson, et al., "Understanding dissociations in dyscalculia: A brain imaging study of the impact of number size on the cerebral networks for exact and approximate calculation," *Brain*, **123,** 2240-2255.

[61] Jeffrey Batholet, "MOOCs: Hype and Hope," *Scientific American*, August 2013, pg 53-61.

[62] Seth Fletcher, "Adaptive Technology: Machine Learning," *Scientific American*, August 2013, pg. 63-68.

[63] Estimated from sales data supplied by the International Federation of Robotics, annual report (2012), using the estimate that the average lifetime of a robot in service is 15 years.

[64] CrowdFlower.com web site.

[65] Meisel (2013) and International Federation of Robotics, Annual Report (2012).

[66] E.g., The Dalberg Report "Impact of the Internet in Africa" 2013.

[67] Brian Patrick (2013) *Selling on Amazon: How you can make a full time income,* GrassRootBook.com. The personal experiences and ideas are useful but the statistics and other background information in

this self-published book are not to be trusted, and therefore the book is not listed in the Bibliography and not recommended.

[68] Comscore, "'Free Shipping Day' Promotion Spurs Late-Season Online Spending Surge, Improving Season-to-Date Growth Rate to 16 Percent vs. Year Ago," comscore.com, Dec. 23, 2012.

[69] Brian Gottlob, PolEcon Research, "Trend Lines: Posted tagged 'retail sales' / E-commerce's small percent of sales has big implications" briangottlob.files, March 21, 2013. Also, Jason Perlow "E-commerce will make the shopping mall a retail wasteland," zdnet.com, January 17, 2013.

[70] Tyler Cowen, "Who Will Prosper in the New World," *New York Times*, August 31, 2013.

[71] There can be no commercial establishments within miles because the land is government land and not for sale. It would be illegal for a business to locate here. It is, of course, illegal for the squatters to locate there, too, and at first the police and bulldozers did try to remove them. This community was well-enough organized to resist them, using a combination of massive peaceful resistance and lawyers, and the authorities seem to have given up. After ten years of continuously occupying the land, the squatters plan to sue the government for title on the basis of their occupation. Similar groups have been successful, e.g., other neighborhoods of Villa San Salvador have attained ownership status using the same process.

[72] Not his real name.

[73] Any time lag between collecting money for products and having to pay for their replacement increases the amount of money circulating in the local economy.

[74] I am a big fan of our space program but I think it is simply not cost-effective to send people on long space trips when it is (will be) so much cheaper to send robots. Space and other planets are much more a natural environment for a robot than for a human. For the very long travel times required for interstellar travel, it will only be possible to send robots, at least for a very, very long time.

---

[75] Irving .J. Good, "Speculations Concerning the First Ultraintelligent Machine," *Advances in Computers*, **6**, 1965.

[76] Verner Vinge, "The Coming Technological Singularity: How to Survive in the Post-Human era," in *Vision-21: Interdisciplinary Science and Engineering in the era of Cyberspace*, G. A. Landis, ed., NASA Publication CP-10129, pp. 11-22 (1993).

[77] This idea was most clearly expressed in a series of science fiction novels by Peter Hamilton starting with *Pandora's Star* (2004) Ballentine Books, N.Y.

[78] Nick Cumming-Bruce, "U.N. Expert Calls for Halt in Military Robot Development," *New York Times*, May 30, 2013.

[79] Data for Figure 8: Global temperature is from NOAA web site, "How Much Has Global Temperature Changed Over the Last 100 Years?" and is the five year running average. Dashed line projection is that predicted by Randers (2012). Speed of fastest computer is extrapolated from data from Kurzweil (2012), same information as in Table B1.

[80] Of course, as noted above, we cannot have any guarantee of this. We really can't yet construct a weather model that takes cloud formation into account, for one thing, and there may be other unknowns in the real world that have not yet been discovered. However, neither do we have any known reason for pessimism at this point.

[81] For an update on the views of Ray Kurzweil, see Kurzweil (2012) and the web site www.KurzweilAI.com. A recent interview with Vernor Vinge is summarized by Barrat (2013) in chapter 8.

[82] World Population to 1950, from Maddison (2006). From 1950 to 2100, from United Nations data (using median projection 2015-2100).

[83] Gross World Product (GWP) per person to 1973, from Maddison (2006). From 2000 to 2012, *CIA World Factbook.* The dashed line in Figure B3 is a conservative extrapolation based on recent data: See text.

---

[84] International Telecommunications Union, press releases, 1997-2012.

[85] Kurzweil (2012), heavy dashed line is best fit for data from 1987-2012, data not shown.

[86] From data and graphs summarized by Kurzweil (2012).

[87] Sources for Figure B4: Fastest computer available worldwide from TOP500 and Kurzweil (2012). In 2013, the worlds' fastest was the Tianhe-2 with a speed of 3.4×10^{16} floating point operations per second (FLOPS). Reported speeds from 1987 to 2013 fit (very well) an exponential growth line with a doubling time of only 1.0 years, so that in the next 20 years there is predicted an increase of 2^{20} = one *million* times (dashed line). Total U.S. non-farm employment from U.S. Bureau of Labor Statistics. Industrial robots installed in U.S. estimated from statistics supplied by International Federation of Robotics (2012 and 2013 reports) with filled circles representing estimated net installations, hollow circle and dashed line representing industry projections.

www.ingramcontent.com/pod-product-compliance
Lightning Source LLC
Chambersburg PA
CBHW071212050326
40689CB00011B/2306

* 9 7 8 0 5 7 8 1 3 0 3 6 1 *